AF521860

The R Reference Manual Base Package Volume 1

The R Development Core Team

Version 1.8.1

A catalogue record for this book is available from the British Library.

First printing, January 2004 (13/1/2004).

Published by Network Theory Limited.

15 Royal Park
Bristol
BS8 3AL
United Kingdom

Email: info@network-theory.co.uk

ISBN 0-9546120-0-0

Further information about this book is available from
`http://www.network-theory.co.uk/R/reference/`

Contents

Publisher's Preface

This reference manual documents the use of R, an environment for statistical computing and graphics.

R is *free software*. The term "free software" refers to your freedom to run, copy, distribute, study, change and improve the software. With R you have all these freedoms.

R is part of the GNU Project. The GNU Project was launched in 1984 to develop a complete Unix-like operating system which is free software: the GNU system. It was conceived as a way of bringing back the cooperative spirit that prevailed in the computing community in earlier days, by removing the obstacles to cooperation imposed by the owners of proprietary software.

You can support the GNU Project by becoming an associate member of the Free Software Foundation. The Free Software Foundation is a tax-exempt charity dedicated to promoting computer users' right to use, study, copy, modify, and redistribute computer programs. It also helps to spread awareness of the ethical and political issues of freedom in the use of software. For more information visit the website `www.fsf.org`.

The development of R itself is guided by the R Foundation, a not for profit organization working in the public interest. Individuals and organisations using R can support its continued development by becoming members of the R Foundation. Further information is available at the website `www.r-project.org`.

Brian Gough
Publisher
November 2003

Chapter 1

The base package

This volume documents the core commands in the *base* package of R. These commands include programming constructs, basic numerical functions and system calls. The base package commands are automatically available when the R environment is started.

The documentation for other base package commands can be found in the second volume "*The R Reference Manual – Base Package – Volume 2*" (ISBN 0-9546120-1-9). These include commands for graphics, random distributions, models, date-time calculations, time-series and example datasets. Documentation for additional packages is available in further volumes of this series.

R is available from many commercial distributors, including the Free Software Foundation on their source code CD-ROMs. Information about R itself can be found online at `www.r-project.org`.

To start the program once it is installed simply use the command `R` on Unix-like systems,

```
$ R
R : Copyright 2003, The R Development Core Team
Type 'demo()' for some demos, 'help()' for on-line help,
or 'help.start()' for a HTML browser interface to help.
Type 'q()' to quit R.
>
```

Commands can then be typed at the R prompt (`>`). The commands given in this manual should generally be entered with arguments in parentheses, e.g. `help()`, as shown in the examples. Typing the name of the function without parentheses, such as `help`, displays its internal definition and does not execute the command.

To obtain online help for any command type `help(`*`command`*`)`. A tutorial for new users of R is available in the book "*An Introduction to R*" (ISBN 0-9541617-4-2).

.Machine *Numerical Characteristics of the Machine*

Description

`.Machine` is a variable holding information on the numerical characteristics of the machine R is running on, such as the largest double or integer and the machine's precision.

Usage

```
.Machine
```

Details

The algorithm is based on Cody's (1988) subroutine MACHAR.

Value

A list with components (for simplicity, the prefix "double" is omitted in the explanations)

`double.eps` the smallest positive floating-point number `x` such that `1 + x != 1`. It equals `base^ulp.digits` if either `base` is 2 or `rounding` is 0; otherwise, it is `(base^ulp.digits) / 2`.

`double.neg.eps`
a small positive floating-point number `x` such that `1 - x != 1`. It equals `base^neg.ulp.digits` if `base` is 2 or `round` is 0; otherwise, it is `(base^neg.ulp.digits) / 2`. As `neg.ulp.digits` is bounded below by `-(digits + 3)`, there might be numbers smaller than `neg.eps` which alter 1 by subtraction.

`double.xmin` the smallest non-vanishing normalized floating-point power of the radix, i.e., `base^min.exp`.

`double.xmax` the largest finite floating-point number. Typically, it is equal to `(1 - neg.eps) * base^max.exp`, but on some machines it is only the second, or perhaps third, largest number, being too small by 1 or 2 units in the last digit of the significand.

`double.base` the radix for the floating-point representation

double.digits the number of base digits in the floating-point significand

double.rounding
the rounding action.
0 if floating-point addition chops;
1 if floating-point addition rounds, but not in the IEEE style;
2 if floating-point addition rounds in the IEEE style;
3 if floating-point addition chops, and there is partial underflow;
4 if floating-point addition rounds, but not in the IEEE style, and there is partial underflow;
5 if floating-point addition rounds in the IEEE style, and there is partial underflow

double.guard the number of guard digits for multiplication with truncating arithmetic. It is 1 if floating-point arithmetic truncates and more than `digits` digits (in base `base`) participate in the post-normalization shift of the floating-point significand in multiplication, and 0 otherwise.

double.ulp.digits
the largest negative integer `i` such that `1 + base^i != 1`, except that it is bounded below by `-(digits + 3)`.

double.neg.ulp.digits
the largest negative integer `i` such that `1 - base^i != 1`, except that it is bounded below by `-(digits + 3)`.

double.exponent
the number of bits (decimal places if `base` is 10) reserved for the representation of the exponent (including the bias or sign) of a floating-point number

double.min.exp
the largest in magnitude negative integer `i` such that `base ^ i` is positive and normalized.

double.max.exp
the smallest positive power of `base` that overflows.

integer.max the largest integer which can be represented.

sizeof.long the number of bytes in a C `long` type.

sizeof.longlong
the number of bytes in a C `long long` type. Will be zero if there is no such type.

`sizeof.longdouble`
: the number of bytes in a C `long double` type. Will be zero if there is no such type.

`sizeof.pointer`
: the number of bytes in a C `SEXP` (s-expression) type.

References

Cody, W. J. (1988) MACHAR: A subroutine to dynamically determine machine parameters. *Transactions on Mathematical Software*, **14**, 4, 303–311.

See Also

`.Platform` for details of the platform.

Examples

```
str(.Machine)
```

`.Platform` *Platform Specific Variables*

Description

`.Platform` is a list with some details of the platform under which R was built. This provides means to write OS portable R code.

Usage

```
.Platform
```

Value

A list with at least the following components:

`OS.type`	character, giving the **O**perating **S**ystem (family) of the computer. One of `"unix"` or `"windows"`.
`file.sep`	character, giving the **file sep**arator, used on your platform, e.g., `"/"` on Unix alikes.
`dynlib.ext`	character, giving the file name **ext**ension of **dyn**amically loadable **lib**raries, e.g., `".dll"` on Windows.
`GUI`	character, giving the type of GUI in use, or `"unknown"` if no GUI can be assumed.
`endian`	character, `"big"` or `"little"`, giving the endianness of the processor in use.

See Also

`R.version` and `Sys.info` give more details about the OS. In particular, `R.version$platform` is the canonical name of the platform under which R was compiled.

`.Machine` for details of the arithmetic used, and `system` for invoking platform-specific system commands.

Examples

```
## Note: this can be done in a system-independent way by
## file.info()$isdir
if(.Platform$OS.type == "unix") {
   system.test <- function(...) {
```

```
    system(paste("test", ...)) == 0
  }
  dir.exists <- function(dir)
    sapply(dir, function(d)system.test("-d", d))
  dir.exists(c(R.home(), "/tmp", "~", "/NO")) # > T T T F
}
```

`.Script` *Scripting Language Interface*

Description

Run a script through its interpreter with given arguments.

Usage

```
.Script(interpreter, script, args, ...)
```

Arguments

`interpreter`	a character string naming the interpreter for the script.
`script`	a character string with the base file name of the script, which must be located in the '`interpreter`' subdirectory of '`R_HOME/share`'.
`args`	a character string giving the arguments to pass to the script.
`...`	further arguments to be passed to `system` when invoking the interpreter on the script.

Note

This function is for R internal use only.

Examples

```
.Script("perl", "massage-Examples.pl",
    paste("tools", system.file("R-ex", package = "tools")))
```

abbreviate *Abbreviate Strings*

Description

Abbreviate strings to at least `minlength` characters, such that they remain *unique* (if they were).

Usage

```
abbreviate(names.arg, minlength = 4, use.classes = TRUE,
           dot = FALSE)
```

Arguments

`names.arg`	a vector of names to be abbreviated.
`minlength`	the minimum length of the abbreviations.
`use.classes`	logical (currently ignored by R).
`dot`	logical; should a dot (`"."`) be appended?

Details

The algorithm used is similar to that of S. First spaces at the beginning of the word are stripped. Then any other spaces are stripped. Next lower case vowels are removed followed by lower case consonants. Finally if the abbreviation is still longer than `minlength` upper case letters are stripped.

Letters are always stripped from the end of the word first. If an element of `names.arg` contains more than one word (words are separated by space) then at least one letter from each word will be retained. If a single string is passed it is abbreviated in the same manner as a vector of strings.

Missing (`NA`) values are not abbreviated.

If `use.classes` is `FALSE` then the only distinction is to be between letters and space. This has NOT been implemented.

Value

A character vector containing abbreviations for the strings in its first argument. Duplicates in the original `names.arg` will be given identical abbreviations. If any non-duplicated elements have the same `minlength` abbreviations then `minlength` is incremented by one and new abbreviations are found for those elements only. This process is repeated until all unique elements of `names.arg` have unique abbreviations.

The character version of `names.arg` is attached to the returned value as a names argument.

See Also

`substr`.

Examples

```
x <- c("abcd", "efgh", "abce")
abbreviate(x, 2)

data(state)
(st.abb <- abbreviate(state.name, 2))
table(nchar(st.abb)) # out of 50, 3 need 4 letters
```

aggregate *Compute Summary Statistics of Data Subsets*

Description

Splits the data into subsets, computes summary statistics for each, and returns the result in a convenient form.

Usage

```
aggregate(x, ...)

## Default S3 method:
aggregate(x, ...)

## S3 method for class 'data.frame':
aggregate(x, by, FUN, ...)

## S3 method for class 'ts':
aggregate(x, nfrequency = 1, FUN = sum, ndeltat = 1,
          ts.eps = getOption("ts.eps"), ...)
```

Arguments

`x`	an R object.
`by`	a list of grouping elements, each as long as the variables in `x`. Names for the grouping variables are provided if they are not given. The elements of the list will be coerced to factors (if they are not already factors).
`FUN`	a scalar function to compute the summary statistics which can be applied to all data subsets.
`nfrequency`	new number of observations per unit of time; must be a divisor of the frequency of `x`.
`ndeltat`	new fraction of the sampling period between successive observations; must be a divisor of the sampling interval of `x`.
`ts.eps`	tolerance used to decide if `nfrequency` is a submultiple of the original frequency.
`...`	further arguments passed to or used by methods.

Details

`aggregate` is a generic function with methods for data frames and time series.

The default method `aggregate.default` uses the time series method if `x` is a time series, and otherwise coerces `x` to a data frame and calls the data frame method.

`aggregate.data.frame` is the data frame method. If `x` is not a data frame, it is coerced to one. Then, each of the variables (columns) in `x` is split into subsets of cases (rows) of identical combinations of the components of `by`, and `FUN` is applied to each such subset with further arguments in `...` passed to it. (I.e., `tapply(VAR, by, FUN, ..., simplify = FALSE)` is done for each variable `VAR` in `x`, conveniently wrapped into one call to `lapply()`.) Empty subsets are removed, and the result is reformatted into a data frame containing the variables in `by` and `x`. The ones arising from `by` contain the unique combinations of grouping values used for determining the subsets, and the ones arising from `x` the corresponding summary statistics for the subset of the respective variables in `x`.

`aggregate.ts` is the time series method. If `x` is not a time series, it is coerced to one. Then, the variables in `x` are split into appropriate blocks of length `frequency(x) / nfrequency`, and `FUN` is applied to each such block, with further (named) arguments in `...` passed to it. The result returned is a time series with frequency `nfrequency` holding the aggregated values.

Author(s)

Kurt Hornik

References

Becker, R. A., Chambers, J. M. and Wilks, A. R. (1988) *The New S Language*. Wadsworth & Brooks/Cole.

See Also

`apply`, `lapply`, `tapply`.

Examples

```
data(state)

## Compute the averages for the variables in 'state.x77',
```

```
## grouped according to the region (Northeast, South, North
## Central, West) that each state belongs to.
aggregate(state.x77, list(Region = state.region), mean)

## Compute the averages according to region and the
## occurrence of more than 130 days of frost.
aggregate(state.x77,
          list(Region = state.region,
               Cold = state.x77[,"Frost"] > 130),
          mean)
## (Note that no state in 'South' is THAT cold.)

data(presidents)
## Compute the average annual approval ratings for American
## presidents.
aggregate(presidents, nf = 1, FUN = mean)
## Give the summer less weight.
aggregate(presidents, nf = 1, FUN = weighted.mean,
          w = c(1, 1, 0.5, 1))
```

agrep *Approximate String Matching (Fuzzy Matching)*

Description

Searches for approximate matches to `pattern` (the first argument) within the string `x` (the second argument) using the Levenshtein edit distance.

Usage

```
agrep(pattern, x, ignore.case = FALSE, value = FALSE,
      max.distance = 0.1)
```

Arguments

`pattern`	a non-empty character string to be matched (*not* a regular expression!)
`x`	character vector where matches are sought.
`ignore.case`	if `FALSE`, the pattern matching is *case sensitive* and if `TRUE`, case is ignored during matching.
`value`	if `FALSE`, a vector containing the (integer) indices of the matches determined is returned and if `TRUE`, a vector containing the matching elements themselves is returned.
`max.distance`	Maximum distance allowed for a match. Expressed either as integer, or as a fraction of the pattern length (will be replaced by the smallest integer not less than the corresponding fraction), or a list with possible components `all`: maximal (overall) distance `insertions`: maximum number/fraction of insertions `deletions`: maximum number/fraction of deletions `substitutions`: maximum number/fraction of substitutions If `all` is missing, it is set to 10%, the other components default to `all`. The component names can be abbreviated.

Details

The Levenshtein edit distance is used as measure of approximateness: it is the total number of insertions, deletions and substitutions required to transform one string into another.

The function is a simple interface to the `apse` library developed by Jarkko Hietaniemi (also used in the Perl String::Approx module).

Value

Either a vector giving the indices of the elements that yielded a match, of, if `value` is `TRUE`, the matched elements.

Author(s)

David Meyer (based on C code by Jarkko Hietaniemi); modifications by Kurt Hornik

See Also

`grep`

Examples

```
agrep("lasy", "1 lazy 2")
agrep("lasy", "1 lazy 2", max = list(sub = 0))
agrep("laysy", c("1 lazy", "1", "1 LAZY"), max = 2)
agrep("laysy", c("1 lazy", "1", "1 LAZY"), max = 2,
      value = TRUE)
agrep("laysy", c("1 lazy", "1", "1 LAZY"), max = 2,
      ignore.case = TRUE)
```

`all` *Are All Values True?*

Description

Given a set of logical vectors, are all of the values true?

Usage

```
all(..., na.rm = FALSE)
```

Arguments

`...`	one or more logical vectors.
`na.rm`	logical. If true `NA` values are removed before the result is computed.

Value

Given a sequence of logical arguments, a logical value indicating whether or not all of the elements of `x` are `TRUE`.

The value returned is `TRUE` if all the values in `x` are `TRUE`, and `FALSE` if any the values in `x` are `FALSE`.

If `x` consists of a mix of `TRUE` and `NA` values, then value is `NA`.

References

Becker, R. A., Chambers, J. M. and Wilks, A. R. (1988) *The New S Language*. Wadsworth & Brooks/Cole.

See Also

`any`, the "complement" of `all`, and `stopifnot(*)` which is an `all(*)` "insurance".

Examples

```
range(x <- sort(round(rnorm(10) - 1.2,1)))
if(all(x < 0)) cat("all x values are negative\n")
```

all.names *Find All Names in an Expression*

Description

Return a character vector containing all the names which occur in an expression or call.

Usage

```
all.names(expr, functions = TRUE, max.names = 200,
          unique = FALSE)

all.vars(expr, functions = FALSE, max.names = 200,
         unique = TRUE)
```

Arguments

`expr`	an expression or call from which the names are to be extracted.
`functions`	a logical value indicating whether function names should be included in the result.
`max.names`	the maximum number of names to be returned.
`unique`	a logical value which indicates whether duplicate names should be removed from the value.

Details

These functions differ only in the default values for their arguments.

Value

A character vector with the extracted names.

Examples

```
all.names(expression(sin(x+y)))
all.vars(expression(sin(x+y)))
```

any *Are Some Values True?*

Description

Given a set of logical vectors, are any of the values true?

Usage

```
any(..., na.rm = FALSE)
```

Arguments

`...`	one or more logical vectors.
`na.rm`	logical. If true `NA` values are removed before the result is computed.

Value

Given a sequence of logical arguments, a logical value indicating whether or not any of the elements of `x` are `TRUE`.

The value returned is `TRUE` if any the values in `x` are `TRUE`, and `FALSE` if all the values in `x` are `FALSE`.

If `x` consists of a mix of `FALSE` and `NA` values, the value is `NA`.

References

Becker, R. A., Chambers, J. M. and Wilks, A. R. (1988) *The New S Language.* Wadsworth & Brooks/Cole.

See Also

`all`, the "complement" of `any`.

Examples

```
range(x <- sort(round(rnorm(10) - 1.2,1)))
if(any(x < 0)) cat("x contains negative values\n")
```

aperm *Array Transposition*

Description

Transpose an array by permuting its dimensions and optionally resizing it.

Usage

```
aperm(a, perm, resize = TRUE)
```

Arguments

`a`	the array to be transposed.
`perm`	the subscript permutation vector, which must be a permutation of the integers `1:n`, where `n` is the number of dimensions of `a`. The default is to reverse the order of the dimensions.
`resize`	a flag indicating whether the vector should be resized as well as having its elements reordered (default `TRUE`).

Value

A transposed version of array `a`, with subscripts permuted as indicated by the array `perm`. If `resize` is `TRUE`, the array is reshaped as well as having its elements permuted, the `dimnames` are also permuted; if `FALSE` then the returned object has the same dimensions as `a`, and the dimnames are dropped.

The function `t` provides a faster and more convenient way of transposing matrices.

Author(s)

Jonathan Rougier did the faster C implementation.

References

Becker, R. A., Chambers, J. M. and Wilks, A. R. (1988) *The New S Language*. Wadsworth & Brooks/Cole.

See Also

`t`, to transpose matrices.

Examples

```
# interchange the first two subscripts on a 3-way array x
x  <- array(1:24, 2:4)
xt <- aperm(x, c(2,1,3))
stopifnot(t(xt[,,2]) == x[,,2],
          t(xt[,,3]) == x[,,3],
          t(xt[,,4]) == x[,,4])
```

append *Vector Merging*

Description

Add elements to a vector.

Usage

```
append(x, values, after=length(x))
```

Arguments

`x`	the vector to be modified.
`values`	to be included in the modified vector.
`after`	a subscript, after which the values are to be appended.

Value

A vector containing the values in `x` with the elements of `values` appended after the specified element of `x`.

References

Becker, R. A., Chambers, J. M. and Wilks, A. R. (1988) *The New S Language*. Wadsworth & Brooks/Cole.

Examples

```
append(1:5, 0:1, after=3)
```

`apply` *Apply Functions Over Array Margins*

Description

Returns a vector or array or list of values obtained by applying a function to margins of an array.

Usage

```
apply(X, MARGIN, FUN, ...)
```

Arguments

`X`	the array to be used.
`MARGIN`	a vector giving the subscripts which the function will be applied over. `1` indicates rows, `2` indicates columns, `c(1,2)` indicates rows and columns.
`FUN`	the function to be applied. In the case of functions like `+`, `%*%`, etc., the function name must be quoted.
`...`	optional arguments to `FUN`.

Details

If `X` is not an array but has a dimension attribute, `apply` attempts to coerce it to an array via `as.matrix` if it is two-dimensional (e.g., data frames) or via `as.array`.

Value

If each call to `FUN` returns a vector of length `n`, then `apply` returns an array of dimension `c(n, dim(X)[MARGIN])` if `n > 1`. If `n` equals `1`, `apply` returns a vector if `MARGIN` has length 1 and an array of dimension `dim(X)[MARGIN]` otherwise. If `n` is `0`, the result has length 0 but not necessarily the "correct" dimension.

If the calls to `FUN` return vectors of different lengths, `apply` returns a list of length `dim(X)[MARGIN]`.

References

Becker, R. A., Chambers, J. M. and Wilks, A. R. (1988) *The New S Language.* Wadsworth & Brooks/Cole.

See Also

`lapply`, `tapply`, and convenience functions `sweep` and `aggregate`.

Examples

```
## Compute row and column sums for a matrix:
x <- cbind(x1 = 3, x2 = c(4:1, 2:5))
dimnames(x)[[1]] <- letters[1:8]
apply(x, 2, mean, trim = .2)
col.sums <- apply(x, 2, sum)
row.sums <- apply(x, 1, sum)
rbind(cbind(x, Rtot = row.sums),
            Ctot = c(col.sums, sum(col.sums)))

stopifnot( apply(x,2, is.vector)) # not ok in R <= 0.63.2

## Sort the columns of a matrix
apply(x, 2, sort)

## function with extra args:
cave <- function(x, c1,c2) c(mean(x[c1]),mean(x[c2]))
apply(x,1, cave,  c1="x1", c2=c("x1","x2"))

ma <- matrix(c(1:4, 1, 6:8), nr = 2)
ma
apply(ma, 1, table)  # a list of length 2
apply(ma, 1, quantile) # 5 x n matrix with rownames

## wasn't ok before R 0.63.1
stopifnot(dim(ma) == dim(apply(ma, 1:2, sum)))
```

`apropos` *Find Objects by (Partial) Name*

Description

`apropos` returns a character vector giving the names of all objects in the search list matching `what`.

`find` is a different user interface to the same task as `apropos`.

Usage

```
apropos(what, where = FALSE, mode = "any")

find(what, mode = "any", numeric. = FALSE,
     simple.words = TRUE)
```

Arguments

`what`	name of an object, or regular expression to match against
`where, numeric.`	a logical indicating whether positions in the search list should also be returned
`mode`	character; if not `"any"`, only objects who's `mode` equals `mode` are searched.
`simple.words`	logical; if `TRUE`, the `what` argument is only searched as whole only word.

Details

If `mode != "any"` only those objects which are of mode `mode` are considered. If `where` is `TRUE`, the positions in the search list are returned as the names attribute.

`find` is a different user interface for the same task as `apropos`. However, by default (`simple.words == TRUE`), only full words are searched.

Author(s)

Kurt Hornik and Martin Maechler (May 1997).

See Also

`objects` for listing objects from one place, `help.search` for searching the help system, `search` for the search path.

Examples

```
apropos("lm")
apropos(ls)
apropos("lq")

lm <- 1:pi
find(lm)          # ".GlobalEnv"   "package:base"
find(lm, num=TRUE) # numbers with these names
find(lm, num=TRUE, mode="function") # only the second one
rm(lm)

apropos(".", mode="list") # a long list

# need a DOUBLE backslash '\\' (in case you don't see
# it anymore)
apropos("\\[")

# everything
length(apropos("."))

# those starting with 'pr'
apropos("^pr")

# the 1-letter things
apropos("^.$")
# the 1-2-letter things
apropos("^..?$")
# the 2-to-4 letter things
apropos("^.{2,4}$")

# the 8-and-more letter things
apropos("^.{8,}$")
table(nchar(apropos("^.{8,}$")))
```

`args` *Argument List of a Function*

Description

Displays the argument names and corresponding default values of a function.

Usage

```
args(name)
```

Arguments

`name`	an interpreted function. If `name` is a character string then the function with that name is found and used.

Details

This function is mainly used interactively. For programming, use `formals` instead.

Value

A function with identical formal argument list but an empty body if given an interpreted function; `NULL` in case of a variable or primitive (non-interpreted) function.

References

Becker, R. A., Chambers, J. M. and Wilks, A. R. (1988) *The New S Language*. Wadsworth & Brooks/Cole.

See Also

`formals`, `help`.

Examples

```
args(c)              # -> NULL (c is a 'primitive' function)
args(plot.default)
```

array *Multi-way Arrays*

Description

Creates or tests for arrays.

Usage

```
array(data = NA, dim = length(data), dimnames = NULL)
as.array(x)
is.array(x)
```

Arguments

`data`	a vector giving data to fill the array.
`dim`	the dim attribute for the array to be created, that is a vector of length one or more giving the maximal indices in each dimension.
`dimnames`	the names for the dimensions. This is a list with one component for each dimension, either NULL or a character vector of the length given by `dim` for that dimension. The list can be names, and the names will be used as names for the dimensions.
`x`	an R object.

Value

`array` returns an array with the extents specified in `dim` and naming information in `dimnames`. The values in `data` are taken to be those in the array with the leftmost subscript moving fastest. If there are too few elements in `data` to fill the array, then the elements in `data` are recycled.

`as.array()` coerces its argument to be an array by attaching a `dim` attribute to it. It also attaches `dimnames` if `x` has `names`. The sole purpose of this is to make it possible to access the `dim`[names] attribute at a later time.

`is.array` returns `TRUE` or `FALSE` depending on whether its argument is an array (i.e., has a `dim` attribute) or not. It is generic: you can write methods to handle specific classes of objects, see InternalMethods.

References

Becker, R. A., Chambers, J. M. and Wilks, A. R. (1988) *The New S Language*. Wadsworth & Brooks/Cole.

See Also

`aperm`, `matrix`, `dim`, `dimnames`.

Examples

```
dim(as.array(letters))
array(1:3, c(2,4)) # recycle 1:3 "2 2/3 times"
#     [,1] [,2] [,3] [,4]
#[1,]    1    3    2    1
#[2,]    2    1    3    2

# funny object:
str(a0 <- array(1:3, 0))
```

as.data.frame *Coerce to a Data Frame*

Description

Functions to check if an object is a data frame, or coerce it if possible.

Usage

```
as.data.frame(x, row.names = NULL, optional = FALSE)
is.data.frame(x)
```

Arguments

`x`	any R object.
`row.names`	`NULL` or a character vector giving the row names for the data frame. Missing values are not allowed.
`optional`	logical. If `TRUE`, setting row names and converting column names (to syntactic names) is optional.

Details

`as.data.frame` is a generic function with many methods, and users and packages can supply further methods.

If a list is supplied, each element is converted to a column in the data frame. Similarly, each column of a matrix is converted separately. This can be overridden if the object has a class which has a method for `as.data.frame`: two examples are matrices of class `"model.matrix"` (which are included as a single column) and list objects of class `"POSIXlt"` which are coerced to class `"POSIXct"`

Character variables are converted to factor columns unless protected by `I`.

If a data frame is supplied, all classes preceding `"data.frame"` are stripped, and the row names are changed if that argument is supplied.

If `row.names = NULL`, row names are constructed from the names or dimnames of `x`, otherwise are the integer sequence starting at one. Few of the methods check for duplicated row names.

Value

`as.data.frame` returns a data frame, normally with all row names `""` if `optional = TRUE`.

`is.data.frame` returns `TRUE` if its argument is a data frame (that is, has `"data.frame"` amongst its classes) and `FALSE` otherwise.

Note

In versions of R prior to 1.4.0 logical columns were converted to factors (as in S3 but not S4).

References

Chambers, J. M. (1992) *Data for models.* Chapter 3 of *Statistical Models in S* eds J. M. Chambers and T. J. Hastie, Wadsworth & Brooks/Cole.

See Also

`data.frame`

`as.environment` *Coerce to an Environment Object*

Description

Converts a number or a character string to the corresponding environment on the search path.

Usage

```
as.environment(object)
```

Arguments

`object`	the object to convert. If it is already an environment, just return it. If it is a number, return the environment corresponding to that position on the search list. If it is a character string, match the string to the names on the search list.

Value

The corresponding environment object.

Author(s)

John Chambers

See Also

`environment` for creation and manipulation, `search`.

Examples

```
as.environment(1) ## the global environment
identical(globalenv(), as.environment(1)) ## is TRUE
## ctest need not be loaded
try(as.environment("package:ctest"))
```

`as.function` *Convert Object to Function*

Description

`as.function` is a generic function which is used to convert objects to functions.

`as.function.default` works on a list `x`, which should contain the concatenation of a formal argument list and an expression or an object of mode `"call"` which will become the function body. The function will be defined in a specified environment, by default that of the caller.

Usage

```
as.function(x, ...)

## Default S3 method:
as.function(x, envir = parent.frame(), ...)
```

Arguments

`x`	object to convert, a list for the default method.
`...`	additional arguments, depending on object
`envir`	environment in which the function should be defined

Value

The desired function.

Author(s)

Peter Dalgaard

See Also

`function`; `alist` which is handy for the construction of argument lists, etc.

Examples

```
as.function(alist(a=,b=2,a+b))
as.function(alist(a=,b=2,a+b))(3)
```

assign *Assign a Value to a Name*

Description

Assign a value to a name in an environment.

Usage

```
assign(x, value, pos = -1, envir = as.environment(pos),
       inherits = FALSE, immediate = TRUE)
```

Arguments

`x`	a variable name (given as a quoted string in the function call).
`value`	a value to be assigned to `x`.
`pos`	where to do the assignment. By default, assigns into the current environment. See the details for other possibilities.
`envir`	the `environment` to use. See the details section.
`inherits`	should the enclosing frames of the environment be inspected?
`immediate`	an ignored compatibility feature.

Details

The `pos` argument can specify the environment in which to assign the object in any of several ways: as an integer (the position in the `search` list); as the character string name of an element in the search list; or as an `environment` (including using `sys.frame` to access the currently active function calls). The `envir` argument is an alternative way to specify an environment, but is primarily there for back compatibility.

`assign` does not dispatch assignment methods, so it cannot be used to set elements of vectors, names, attributes, etc.

Note that assignment to an attached list or data frame changes the attached copy and not the original object: see `attach`.

Value

This function is invoked for its side effect, which is assigning `value` to the variable `x`. If no `envir` is specified, then the assignment takes place in the currently active environment.

If `inherits` is `TRUE`, enclosing environments of the supplied environment are searched until the variable `x` is encountered. The value is then assigned in the environment in which the variable is encountered. If the symbol is not encountered then assignment takes place in the user's workspace (the global environment).

If `inherits` is `FALSE`, assignment takes place in the initial frame of `envir`.

References

Becker, R. A., Chambers, J. M. and Wilks, A. R. (1988) *The New S Language*. Wadsworth & Brooks/Cole.

See Also

`<-`, `get`, `exists`, `environment`.

Examples

```
for(i in 1:6) { # Create objects  'r1', 'r2', ... 'r6' --
 nam <- paste("r",i, sep=".")
 assign(nam, 1:i)
}
ls(pat="^r..$")

## Global assignment within a function:
myf <- function(x) {
 innerf <- function(x)
   assign("Global.res", x^2, env = .GlobalEnv)
 innerf(x+1)
}
myf(3)
Global.res # 16

a <- 1:4
assign("a[1]", 2)
a[1] == 2          # FALSE
get("a[1]") == 2   # TRUE
```

assignOps *Assignment Operators*

Description

Assign a value to a name.

Usage

```
x <- value
x <<- value
value -> x
value ->> x

x = value
```

Arguments

`x`	a variable name (possibly quoted).
`value`	a value to be assigned to `x`.

Details

There are three different assignment operators: two of them have leftwards and rightwards forms.

The operators `<-` and `=` assign into the environment in which they are evaluated. The `<-` can be used anywhere, but the `=` is only allowed at the top level (that is, in the complete expression typed by the user) or as one of the subexpressions in a braced list of expressions.

The operators `<<-` and `->>` cause a search to made through the environment for an existing definition of the variable being assigned. If such a variable is found then its value is redefined, otherwise assignment takes place globally. Note that their semantics differ from that in the S language, but is useful in conjunction with the scoping rules of R.

In all the assignment operator expressions, `x` can be a name or an expression defining a part of an object to be replaced (e.g., `z[[1]]`). The name does not need to be quoted, though it can be.

The leftwards forms of assignment `<- = <<-` group right to left, the other from left to right.

Value

`value`. Thus one can use `a <- b <- c <- 6`.

References

Becker, R. A., Chambers, J. M. and Wilks, A. R. (1988) *The New S Language*. Wadsworth & Brooks/Cole.

Chamber, J. M. (1998) *Programming with Data. A Guide to the S Language*. Springer (for =).

See Also

`assign`, `environment`.

`attach` *Attach Set of R Objects to Search Path*

Description

The database is attached to the R search path. This means that the database is searched by R when evaluating a variable, so objects in the database can be accessed by simply giving their names.

Usage

```
attach(what, pos = 2, name = deparse(substitute(what)))
```

Arguments

`what`	"database". This may currently be a `data.frame` or `list` or a R data file created with `save`.
`pos`	integer specifying position in `search()` where to attach.
`name`	alternative way to specify the database to be attached.

Details

When evaluating a variable or function name R searches for that name in the databases listed by `search`. The first name of the appropriate type is used.

By attaching a data frame to the search path it is possible to refer to the variables in the data frame by their names alone, rather than as components of the data frame (e.g. `name` rather than `frame$name`).

By default the database is attached in position 2 in the search path, immediately after the user's workspace and before all previously loaded packages and previously attached databases. This can be altered to attach later in the search path with the `pos` option, but you cannot attach at `pos=1`.

Note that by default assignment is not performed in an attached database. Attempting to modify a variable or function in an attached database will actually create a modified version in the user's workspace (the R global environment). If you use `assign` to assign to an attached list or data frame, you only alter the attached copy, not the original object. For this reason `attach` can lead to confusion.

Value

The `environment` is returned invisibly with a `"name"` attribute.

References

Becker, R. A., Chambers, J. M. and Wilks, A. R. (1988) *The New S Language.* Wadsworth & Brooks/Cole.

See Also

`library`, `detach`, `search`, `objects`, `environment`, `with`.

Examples

```
data(women)
summary(women$height) ## refers to variable 'height' in the
                      ## data frame
attach(women)
summary(height)      ## The same variable now available
                     ## by name

height <- height*2.54  ## Don't do this. It creates a new
                       ## variable
detach("women")
summary(height)      ## The new variable created by
                     ## modifying 'height'
rm(height)
```

attr *Object Attributes*

Description

Get or set specific attributes of an object.

Usage

```
attr(x, which)
attr(x, which) <- value
```

Arguments

`x`	an object whose attributes are to be accessed.
`which`	a character string specifying which attribute is to be accessed.
`value`	an object, the new value of the attribute.

Value

This function provides access to a single object attribute. The simple form above returns the value of the named attribute. The assignment form causes the named attribute to take the value on the right of the assignment symbol.

The first form first looks for an exact match to `code` amongst the attributed of `x`, then a partial match. If no exact match is found and more than one partial match is found, the result is `NULL`.

References

Becker, R. A., Chambers, J. M. and Wilks, A. R. (1988) *The New S Language.* Wadsworth & Brooks/Cole.

See Also

`attributes`

Examples

```
# create a 2 by 5 matrix
x <- 1:10
attr(x,"dim") <- c(2, 5)
```

`attributes` *Object Attribute Lists*

Description

These functions access an object's attribute list. The first form above returns the an object's attribute list. The assignment forms make the list on the right-hand side of the assignment the object's attribute list (if appropriate).

Usage

```
attributes(obj)
attributes(obj) <- value
mostattributes(obj) <- value
```

Arguments

`obj`	an object
`value`	an appropriate attribute list, or `NULL`.

Details

The `mostattributes` assignment takes special care for the `dim`, `names` and `dimnames` attributes, and assigns them only when that is valid whereas as `attributes` assignment would give an error in that case.

References

Becker, R. A., Chambers, J. M. and Wilks, A. R. (1988) *The New S Language.* Wadsworth & Brooks/Cole.

See Also

`attr`.

Examples

```
x <- cbind(a=1:3, pi=pi) # simple matrix w/ dimnames
str(attributes(x))

## strip an object's attributes:
attributes(x) <- NULL
```

```
x # now just a vector of length 6

mostattributes(x) <-
  list(mycomment = "really special", dim = 3:2,
       dimnames = list(LETTERS[1:3], letters[1:5]),
       names = paste(1:6))
x # dim(), but not {dim}names
```

autoload *On-demand Loading of Packages*

Description

autoload creates a promise-to-evaluate **autoloader** and stores it with name **name** in **.AutoloadEnv** environment. When R attempts to evaluate **name**, **autoloader** is run, the package is loaded and **name** is re-evaluated in the new package's environment. The result is that R behaves as if **file** was loaded but it does not occupy memory.

Usage

```
autoload(name, package, ...)
autoloader(name, package, ...)
.AutoloadEnv
```

Arguments

name	string giving the name of an object.
package	string giving the name of a package containing the object.
...	other arguments to **library**.

Value

This function is invoked for its side-effect. It has no return value as of R 1.7.0.

See Also

delay, **library**

Examples

```
autoload("line","eda")
search()
ls("Autoloads")

data(cars)
plot(cars)
z<-line(cars)
abline(coef(z))
```

```
search()
detach("package:eda")
search()
z<-line(cars)
search()
```

ave *Group Averages Over Level Combinations of Factors*

Description

Subsets of `x[]` are averaged, where each subset consist of those observations with the same factor levels.

Usage

```
ave(x, ..., FUN = mean)
```

Arguments

`x`	A numeric.
`...`	Grouping variables, typically factors, all of the same length as `x`.
`FUN`	Function to apply for each factor level combination.

Value

A numeric vector, say `y` of length `length(x)`. If `...` is `g1,g2`, e.g., `y[i]` is equal to `FUN(x[j]`, for all `j` with `g1[j]==g1[i]` and `g2[j]==g2[i])`.

See Also

`mean`, `median`.

Examples

```
ave(1:3) # no grouping -> grand mean

data(warpbreaks)
attach(warpbreaks)
ave(breaks, wool)
ave(breaks, tension)
ave(breaks, tension, FUN = function(x)mean(x, trim=.1))
plot(breaks, main =
     "ave(Warpbreaks) for wool x tension combinations")
lines(ave(breaks, wool, tension), type='s', col = "blue")
lines(ave(breaks, wool, tension, FUN=median),
      type='s', col = "green")
legend(40,70, c("mean","median"), lty=1,
```

```
        col=c("blue","green"), bg="gray90")
  detach()
```

basename *Manipulate File Paths*

Description

basename removes all of the path up to the last path separator (if any).

dirname returns the part of the path up to (but excluding) the last path separator, or "." if there is no path separator.

Usage

```
basename(path)
dirname(path)
```

Arguments

path	character vector, containing path names.

Details

For dirname tilde expansion is done: see the description of path.expand.

Trailing file separators are removed before dissecting the path, and for dirname any trailing file separators are removed from the result.

Value

A character vector of the same length as path. A zero-length input will give a zero-length output with no error (unlike R < 1.7.0).

See Also

file.path, path.expand.

Examples

```
basename(file.path("","p1","p2","p3", c("file1", "file2")))
dirname(file.path("","p1","p2","p3","filename"))
```

BATCH *Batch Execution of R*

Description

Run R non-interactively with input from a given file and place output (stdout/stderr) to another file.

Usage

```
R CMD BATCH [options] infile [outfile]
```

Arguments

`infile`	the name of a file with R code to be executed.
`options`	a list of R command line options, e.g., for setting the amount of memory available and controlling the load/save process. If `infile` starts with a `-`, use `--` as the final option.
`outfile`	the name of a file to which to write output. If not given, the name used is the one of `infile`, with a possible '`.R`' extension stripped, and '`.Rout`' appended.

Details

By default, the input commands are printed along with the output. To suppress this behavior, add `options(echo = FALSE)` at the beginning of `infile`.

The `infile` can have end of line marked by LF or CRLF (but not just CR), and files with a missing EOL mark are processed correctly.

Using `R CMD BATCH` sets the GUI to `"none"`, so none of `x11`, `jpeg` and `png` are available.

Note

Unlike `Splus BATCH`, this does not run the R process in the background. In most shells, `R CMD BATCH [options] infile [outfile] &` will do so.

bindenv *Binding and Environment Adjustments*

Description

These functions represent an experimental interface for adjustments to environments and bindings within environments. They allow for locking environments as well as individual bindings, and for linking a variable to a function.

Usage

```
lockEnvironment(env, bindings = FALSE)
environmentIsLocked(env)
lockBinding(sym, env)
unlockBinding(sym, env)
bindingIsLocked(sym, env)
makeActiveBinding(sym, fun, env)
bindingIsActive(sym, env)
```

Arguments

`env`	an environment.
`bindings`	logical specifying whether bindings should be locked.
`sym`	a name object or character string
`fun`	a function taking zero or one arguments

Details

The function `lockEnvironment` locks its environment argument, which must be a proper environment, not NULL. Locking the NULL (base) environment may be supported later. Locking the environment prevents adding or removing variable bindings from the environment. Changing the value of a variable is still possible unless the binding has been locked.

`lockBinding` locks individual bindings in the specified environment. The value of a locked binding cannot be changed. Locked bindings may be removed from an environment unless the environment is locked.

`makeActiveBinding` installs `fun` so that getting the value of `sym` calls `fun` with no arguments, and assigning to `sym` calls `fun` with one argument, the value to be assigned. This allows things like C variables linked to R variables and variables linked to data bases to be implemented. It may also be useful for making thread-safe versions of some system globals.

Author(s)

Luke Tierney

Examples

```
# locking environments
e<-new.env()
assign("x",1, env=e)
get("x",env=e)
lockEnvironment(e)
get("x",env=e)
assign("x",2, env=e)
try(assign("y",2, env=e)) # error

# locking bindings
e<-new.env()
assign("x",1, env=e)
get("x",env=e)
lockBinding("x", e)
try(assign("x",2, env=e)) # error
unlockBinding("x", e)
assign("x",2, env=e)
get("x",env=e)

# active bindings
f<-local({
    x <- 1
    function(v) {
       if (missing(v))
           cat("get\n")
       else {
           cat("set\n")
           x <<- v
       }
       x
    }
})
makeActiveBinding("fred", f, .GlobalEnv)
```

```
bindingIsActive("fred", .GlobalEnv)
fred
fred<-2
fred
```

body *Access to and Manipulation of the Body of a Function*

Description

Get or set the body of a function.

Usage

```
body(fun = sys.function(sys.parent()))
body(fun, envir = parent.frame()) <- value
```

Arguments

`fun`	a function object, or see Details.
`envir`	environment in which the function should be defined.
`value`	an expression or a list of R expressions.

Details

For the first form, `fun` can be a character string naming the function to be manipulated, which is searched for from the parent environment. If it is not specified, the function calling `body` is used.

Value

`body` returns the body of the function specified.

The assignment form sets the body of a function to the list on the right hand side.

See Also

`alist`, `args`, `function`.

Examples

```
body(body)
f <- function(x) x^5
body(f) <- expression(5^x)
## or equivalently  body(f) <- list(quote(5^x))
f(3) # = 125
str(body(f))
```

bquote *Partial substitution in expressions*

Description

An analogue of the LISP backquote macro. `bquote` quotes its argument except that terms wrapped in `.()` are evaluated in the specified `where` environment.

Usage

```
bquote(expr, where = parent.frame())
```

Arguments

`expr`	An expression
`where`	An environment

Value

An expression

See Also

`quote`, `substitute`

Examples

```
a<-2

bquote(a==a)
quote(a==a)

bquote(a==.(a))
substitute(a==A, list(A=a))

plot(1:10,a*(1:10), main=bquote(a==.(a)))
```

browseEnv *Browse Objects in Environment*

Description

The `browseEnv` function opens a browser with list of objects currently in `sys.frame()` environment.

Usage

```
browseEnv(envir = .GlobalEnv, pattern,
          excludepatt = "^last\\.warning",
          html = .Platform$OS.type != "mac",
          expanded = TRUE, properties = NULL,
          main = NULL, debugMe = FALSE)
```

Arguments

`envir`	an `environment` the objects of which are to be browsed.
`pattern`	a regular expression for object subselection is passed to the internal `ls()` call.
`excludepatt`	a regular expression for *dropping* objects with matching names.
`html`	is used on non Macintosh machines to display the workspace on a HTML page in your favorite browser.
`expanded`	whether to show one level of recursion. It can be useful to switch it to `FALSE` if your workspace is large. This option is ignored if `html` is set to `FALSE`.
`properties`	a named list of global properties (of the objects chosen) to be showed in the browser; when `NULL` (as per default), user, date, and machine information is used.
`main`	a title string to be used in the browser; when `NULL` (as per default) a title is constructed.
`debugMe`	logical switch; if true, some diagnostic output is produced.

Details

Very experimental code. Only allows one level of recursion into object structures. The HTML version is not dynamic.

It can be generalized. See the source file '`databrowser.R`' in the R distribution for details.

`wsbrowser()` is currently just an internally used function; its argument list will certainly change.

Most probably, this should rather work through using the '`tkWidget`' package (from `www.Bioconductor.org`).

See Also

`str`, `ls`.

Examples

```
if(interactive()) {
   ## create some interesting objects :
   ofa <- ordered(4:1)
   ex1 <- expression(1+ 0:9)
   ex3 <- expression(u,v, 1+ 0:9)
   example(factor, echo = FALSE)
   example(table, echo = FALSE)
   example(ftable, echo = FALSE)
   example(lm, echo = FALSE)
   example(str, echo = FALSE)

   ## and browse them:
   browseEnv()

   ## a (simple) function's environment:
   af12 <- approxfun(1:2, 1:2, method = "const")
   browseEnv(envir = environment(af12))
 }
```

browser *Environment Browser*

Description

Interrupt the execution of an expression and allow the inspection of the environment where `browser` was called from.

Usage

```
browser()
```

Details

A call to `browser` causes a pause in the execution of the current expression and runs a copy of the R interpreter which has access to variables local to the environment where the call took place.

Local variables can be listed with `ls`, and manipulated with R expressions typed to this sub-interpreter. The sub-interpreter can be exited by typing `c`. Execution then resumes at the statement following the call to `browser`.

Typing `n` causes the step-through-debugger, to start and it is possible to step through the remainder of the function one line at a time.

Typing `Q` quits the current execution and returns you to the top-level prompt.

References

Becker, R. A., Chambers, J. M. and Wilks, A. R. (1988) *The New S Language.* Wadsworth & Brooks/Cole.

Chambers, J. M. (1998) *Programming with Data. A Guide to the S Language.* Springer.

See Also

`debug`, and `traceback` for the stack on error.

`browseURL` *Load URL into a WWW Browser*

Description

Load a given URL into a WWW browser.

Usage

```
browseURL(url, browser = getOption("browser"))
```

Arguments

`url`	a non-empty character string giving the URL to be loaded.
`browser`	a non-empty character string giving the name of the program to be used as hypertext browser. It should be in the PATH, or a full path specified.

Details

If `browser` supports remote control and R knows how to perform it, the URL is opened in any already running browser or a new one if necessary. This mechanism currently is available for browsers which support the `"-remote openURL(...)"` interface (which includes Netscape 4.x, 6.2.x (but not 6.0/1), Opera 5/6 and Mozilla $\geq$ 0.9.5), Galeon, KDE konqueror (via kfmclient) and the GNOME interface to Mozilla. Netscape 7.0 behaves slightly differently, and you will need to open it first. Note that the type of browser is determined from its name, so this mechanism will only be used if the browser is installed under its canonical name.

Because `"-remote"` will use any browser displaying on the X server (whatever machine it is running on), the remote control mechanism is only used if `DISPLAY` points to the local host. This may not allow displaying more than one URL at a time from a remote host.

bug.report *Send a Bug Report*

Description

Invokes an editor to write a bug report and optionally mail it to the automated r-bugs repository at `r-bugs@r-project.org`. Some standard information on the current version and configuration of R are included automatically.

Usage

```
bug.report(subject = "", ccaddress = Sys.getenv("USER"),
           method = getOption("mailer"),
           address = "r-bugs@r-project.org",
           file = "R.bug.report")
```

Arguments

`subject`	Subject of the email. Please do not use single quotes (`'`) in the subject! File separate bug reports for multiple bugs
`ccaddress`	Optional email address for copies (default is current user). Use `ccaddress = FALSE` for no copies.
`method`	Submission method, one of `"mailx"`, `"gnudoit"`, `"none"`, or `"ess"`.
`address`	Recipient's email address.
`file`	File to use for setting up the email (or storing it when method is `"none"` or sending mail fails).

Details

Currently direct submission of bug reports works only on Unix systems. If the submission method is `"mailx"`, then the default editor is used to write the bug report. Which editor is used can be controlled using `options`, type `getOption("editor")` to see what editor is currently defined. Please use the help pages of the respective editor for details of usage. After saving the bug report (in the temporary file opened) and exiting the editor the report is mailed using a Unix command line mail utility such as `mailx`. A copy of the mail is sent to the current user.

If method is `"gnudoit"`, then an emacs mail buffer is opened and used for sending the email.

If method is `"none"` or `NULL` (which is the default on Windows systems), then only an editor is opened to help writing the bug report. The report can then be copied to your favorite email program and be sent to the r-bugs list.

If method is `"ess"` the body of the mail is simply sent to stdout.

Value

Nothing useful.

When is there a bug?

If R executes an illegal instruction, or dies with an operating system error message that indicates a problem in the program (as opposed to something like "disk full"), then it is certainly a bug.

Taking forever to complete a command can be a bug, but you must make certain that it was really R's fault. Some commands simply take a long time. If the input was such that you KNOW it should have been processed quickly, report a bug. If you don't know whether the command should take a long time, find out by looking in the manual or by asking for assistance.

If a command you are familiar with causes an R error message in a case where its usual definition ought to be reasonable, it is probably a bug. If a command does the wrong thing, that is a bug. But be sure you know for certain what it ought to have done. If you aren't familiar with the command, or don't know for certain how the command is supposed to work, then it might actually be working right. Rather than jumping to conclusions, show the problem to someone who knows for certain.

Finally, a command's intended definition may not be best for statistical analysis. This is a very important sort of problem, but it is also a matter of judgment. Also, it is easy to come to such a conclusion out of ignorance of some of the existing features. It is probably best not to complain about such a problem until you have checked the documentation in the usual ways, feel confident that you understand it, and know for certain that what you want is not available. The mailing list `r-devel@r-project.org` is a better place for discussions of this sort than the bug list.

If you are not sure what the command is supposed to do after a careful reading of the manual this indicates a bug in the manual. The manual's job is to make everything clear. It is just as important to report documentation bugs as program bugs.

If the online argument list of a function disagrees with the manual, one of them must be wrong, so report the bug.

How to report a bug

When you decide that there is a bug, it is important to report it and to report it in a way which is useful. What is most useful is an exact description of what commands you type, from when you start R until the problem happens. Always include the version of R, machine, and operating system that you are using; type `version` in R to print this. To help us keep track of which bugs have been fixed and which are still open please send a separate report for each bug.

The most important principle in reporting a bug is to report FACTS, not hypotheses or categorizations. It is always easier to report the facts, but people seem to prefer to strain to posit explanations and report them instead. If the explanations are based on guesses about how R is implemented, they will be useless; we will have to try to figure out what the facts must have been to lead to such speculations. Sometimes this is impossible. But in any case, it is unnecessary work for us.

For example, suppose that on a data set which you know to be quite large the command `data.frame(x, y, z, monday, tuesday)` never returns. Do not report that `data.frame()` fails for large data sets. Perhaps it fails when a variable name is a day of the week. If this is so then when we got your report we would try out the `data.frame()` command on a large data set, probably with no day of the week variable name, and not see any problem. There is no way in the world that we could guess that we should try a day of the week variable name.

Or perhaps the command fails because the last command you used was a `[` method that had a bug causing R's internal data structures to be corrupted and making the `data.frame()` command fail from then on. This is why we need to know what other commands you have typed (or read from your startup file).

It is very useful to try and find simple examples that produce apparently the same bug, and somewhat useful to find simple examples that might be expected to produce the bug but actually do not. If you want to debug the problem and find exactly what caused it, that is wonderful. You should still report the facts as well as any explanations or solutions.

Invoking R with the '`--vanilla`' option may help in isolating a bug. This ensures that the site profile and saved data files are not read.

A bug report can be generated using the `bug.report()` function. This automatically includes the version information and sends the bug to the correct address. Alternatively the bug report can be emailed to `r-bugs@r-project.org` or submitted to the Web page at `http://bugs.r-project.org`.

Bug reports on **contributed packages** should perhaps be sent to the package maintainer rather than to r-bugs.

Author(s)

This help page is adapted from the Emacs manual and the R FAQ

See Also

R FAQ

`builtins` *Returns the names of all built-in objects*

Description

Return the names of all the built-in objects. These are fetched directly from the symbol table of the R interpreter.

Usage

```
builtins(internal = FALSE)
```

Arguments

`internal`	a logical indicating whether only "internal" functions (which can be called via `.Internal`) should be returned.

by *Apply a Function to a Data Frame split by Factors*

Description

Function `by` is an object-oriented wrapper for `tapply` applied to data frames.

Usage

```
by(data, INDICES, FUN, ...)
```

Arguments

`data`	an R object, normally a data frame, possibly a matrix.
`INDICES`	a factor or a list of factors, each of length `nrow(x)`.
`FUN`	a function to be applied to data frame subsets of `x`.
`...`	further arguments to `FUN`.

Details

A data frame is split by row into data frames subsetted by the values of one or more factors, and function `FUN` is applied to each subset in turn.

Object `data` will be coerced to a data frame by default.

Value

A list of class `"by"`, giving the results for each subset.

See Also

`tapply`

Examples

```
data(warpbreaks)
attach(warpbreaks)
by(warpbreaks[, 1:2], tension, summary)
by(warpbreaks[, 1], list(wool=wool, tension=tension),
   summary)
by(warpbreaks, tension,
   function(x) lm(breaks ~ wool, data=x))
```

```
## now suppose we want to extract the coefficients by group
tmp <- by(warpbreaks, tension,
          function(x) lm(breaks ~ wool, data=x))
sapply(tmp, coef)

detach("warpbreaks")
```

c *Combine Values into a Vector or List*

Description

This is a generic function which combines its arguments.

The default method combines its arguments to form a vector. All arguments are coerced to a common type which is the type of the returned value.

Usage

```
c(..., recursive=FALSE)
```

Arguments

`...`	objects to be concatenated.
`recursive`	logical. If `recursive=TRUE`, the function recursively descends through lists combining all their elements into a vector.

References

Becker, R. A., Chambers, J. M. and Wilks, A. R. (1988) *The New S Language*. Wadsworth & Brooks/Cole.

See Also

`unlist` and `as.vector` to produce attribute-free vectors.

Examples

```
c(1,7:9)
c(1:5, 10.5, "next")

## append to a list:
ll <- list(A = 1, c="C")
## do *not* use
c(ll, d = 1:3) # which is == c(ll, as.list(c(d=1:3))
## but rather
c(ll, d = list(1:3)) # c() combining two lists

c(list(A=c(B=1)), recursive=TRUE)
```

```
c(options(), recursive=TRUE)
c(list(A=c(B=1,C=2), B=c(E=7)), recursive=TRUE)
```

`call` *Function Calls*

Description

Create or test for objects of mode `"call"`.

Usage

```
call(name, ...)
is.call(x)
as.call(x)
```

Arguments

`name`	a character string naming the function to be called.
`...`	arguments to be part of the call.
`x`	an arbitrary R object.

Details

`call` returns an unevaluated function call, that is, an unevaluated expression which consists of the named function applied to the given arguments (`name` must be a quoted string which gives the name of a function to be called).

`is.call` is used to determine whether `x` is a call (i.e., of mode `"call"`). It is generic: you can write methods to handle specific classes of objects, see InternalMethods.

Objects of mode `"list"` can be coerced to mode `"call"`. The first element of the list becomes the function part of the call, so should be a function or the name of one (as a symbol; a quoted string will not do).

References

Becker, R. A., Chambers, J. M. and Wilks, A. R. (1988) *The New S Language*. Wadsworth & Brooks/Cole.

See Also

`do.call` for calling a function by name and argument list; `Recall` for recursive calling of functions; further `is.language`, `expression`, `function`.

Examples

```
is.call(call) # FALSE: Functions are NOT calls

# set up a function call to round with argument 10.5
cl <- call("round", 10.5)
is.call(cl) # TRUE
cl
# such a call can also be evaluated.
eval(cl) # [1] 10
```

`capabilities` *Report Capabilities of this Build of R*

Description

Report on the optional features which have been compiled into this build of R.

Usage

```
capabilities(what = NULL)
```

Arguments

`what`	character vector or `NULL`, specifying required components. `NULL` implies that all are required.

Value

A named logical vector. Current components are

`jpeg`	Is the `jpeg` function operational?
`png`	Is the png function operational?
`tcltk`	Is the **tcltk** package operational?
`X11`	(Unix) Are `X11` and the data editor available?
`GNOME`	(Unix) Is the GNOME GUI in use and are `GTK` and `GNOME` graphics devices available?
`libz`	Is `gzfile` available? From R 1.5.0 this will always be true.
`http/ftp`	Are `url` and the internal method for `download.file` available?
`sockets`	Are `make.socket` and related functions available?
`libxml`	Is there support for integrating `libxml` with the R event loop?
`cledit`	Is command-line editing available in the current R session? This is false in non-interactive sessions. It will be true if `readline` supported has been compiled in and '`--no-readline`' was *not* invoked.
`IEEE754`	Does this platform have IEEE 754 arithmetic? Note that this is more correctly known by the international standard IEC 60559.

`bzip2` Is `bzfile` available?

`PCRE` Is the Perl-Compatible Regular Expression library available? This is needed for the `perl = TRUE` option to `grep` are related function.

See Also

`.Platform`

Examples

```
capabilities()

if(!capabilities("http/ftp"))
   warning("internal download.file() is not available")

## See also the examples for 'connections'.
```

capture.output *Send output to a character string or file*

Description

Evaluates its arguments with the output being returned as a character string or sent to a file. Related to `sink` in the same way that `with` is related to `attach`.

Usage

```
capture.output(..., file = NULL, append = FALSE)
```

Arguments

`...`	Expressions to be evaluated
`file`	A file name or a connection, or `NULL` to return the output as a string. If the connection is not open it will be opened and then closed on exit.
`append`	Append or overwrite the file?

Value

A character string, or `NULL` if a `file` argument was supplied.

See Also

`sink`, `textConnection`

Examples

```
glmout<-capture.output(example(glm))
glmout[1:5]
capture.output(1+1,2+2)
capture.output({1+1;2+2})

## on Unix with enscript available
ps<-pipe("enscript -o tempout.ps","w")
capture.output(example(glm), file=ps)
close(ps)
```

cat *Concatenate and Print*

Description

Prints the arguments, coercing them if necessary to character mode first.

Usage

```
cat(... , file = "", sep = " ", fill = FALSE, labels = NULL,
    append = FALSE)
```

Arguments

`...`	R objects which are coerced to character strings, concatenated, and printed, with the remaining arguments controlling the output.
`file`	A connection, or a character string naming the file to print to. If `""` (the default), `cat` prints to the standard output connection, the console unless redirected by `sink`. If it is `"|cmd"`, the output is piped to the command given by '`cmd`', by opening a pipe connection.
`sep`	character string to insert between the objects to print.
`fill`	a logical or numeric controlling how the output is broken into successive lines. If `FALSE` (default), only newlines created explicitly by '`\n`' are printed. Otherwise, the output is broken into lines with print width equal to the option `width` if `fill` is `TRUE`, or the value of `fill` if this is numeric.
`labels`	character vector of labels for the lines printed. Ignored if `fill` is `FALSE`.
`append`	logical. Only used if the argument `file` is the name of file (and not a connection or `"|cmd"`). If `TRUE` output will be appended to `file`; otherwise, it will overwrite the contents of `file`.

Details

`cat` converts its arguments to character strings, concatenates them, separating them by the given `sep=` string, and then prints them.

No linefeeds are printed unless explicitly requested by '\n' or if generated by filling (if argument `fill` is `TRUE` or numeric.)

`cat` is useful for producing output in user-defined functions.

Value

None (invisible `NULL`).

References

Becker, R. A., Chambers, J. M. and Wilks, A. R. (1988) *The New S Language.* Wadsworth & Brooks/Cole.

See Also

`print`, `format`, and `paste` which concatenates into a string.

Examples

```
iter <- rpois(1, lambda=10)
## print an informative message
cat("iteration = ", iter <- iter + 1, "\n")

## 'fill' and label lines:
cat(paste(letters, 100* 1:26), fill = TRUE,
    labels = paste("{",1:10,"}:",sep=""))
```

cbind *Combine R Objects by Rows or Columns*

Description

Take a sequence of vector, matrix or data frames arguments and combine by *c*olumns or *r*ows, respectively. These are generic functions with methods for other R classes.

Usage

```
cbind(..., deparse.level = 1)
rbind(..., deparse.level = 1)
```

Arguments

`...`	vectors or matrices. These can be given as named arguments.
`deparse.level`	integer controlling the construction of labels; currently, `1` is the only possible value.

Details

The functions `cbind` and `rbind` are generic, with methods for data frames. The data frame method will be used if an argument is a data frame and the rest are vectors or matrices. There can be other methods; in particular, there is one for time series objects.

The `rbind` data frame method takes the classes of the columns from the first data frame. Factors have their levels expanded as necessary (in the order of the levels of the levelsets of the factors encountered) and the result is an ordered factor if and only if all the components were ordered factors. (The last point differs from S-PLUS.)

If there are several matrix arguments, they must all have the same number of columns (or rows) and this will be the number of columns (or rows) of the result. If all the arguments are vectors, the number of columns (rows) in the result is equal to the length of the longest vector. Values in shorter arguments are recycled to achieve this length (with a `warning` if they are recycled only *fractionally*).

When the arguments consist of a mix of matrices and vectors the number of columns (rows) of the result is determined by the number of columns (rows) of the matrix arguments. Any vectors have their values recycled or subsetted to achieve this length.

For `cbind` (`rbind`), vectors of zero length (including `NULL`) are ignored unless the result would have zero rows (columns), for S compatibility. (Zero-extent matrices do not occur in S3 and are not ignored in R.)

Value

A matrix or data frame combining the `...` arguments column-wise or row-wise.

For `cbind` (`rbind`) the column (row) names are taken from the names of the arguments, or where those are not supplied by deparsing the expressions given (if that gives a sensible name). The names will depend on whether data frames are included: see the examples.

Note

The method dispatching is *not* done via `UseMethod()`, but by C-internal dispatching. Therefore, there is no need for, e.g., `rbind.default`.

The dispatch algorithm is described in the source file ('`.../src/main/bind.c`') as

1. For each argument we get the list of possible class memberships from the class attribute.
2. We inspect each class in turn to see if there is an an applicable method.
3. If we find an applicable method we make sure that it is identical to any method determined for prior arguments. If it is identical, we proceed, otherwise we immediately drop through to the default code.

If you want to combine other objects with data frames, it may be necessary to coerce them to data frames first. (Note that this algorithm can result in calling the data frame method if the arguments are all either data frames or vectors, and this will result in the coercion of character vectors to factors.)

References

Becker, R. A., Chambers, J. M. and Wilks, A. R. (1988) *The New S Language.* Wadsworth & Brooks/Cole.

See Also

`c` to combine vectors (and lists) as vectors, `data.frame` to combine vectors and matrices as a data frame.

Examples

```
cbind(1, 1:7) # the '1' (= shorter vector) is recycled
cbind(1:7, diag(3)) # vector is subset -> warning

cbind(0, rbind(1, 1:3))
cbind(I=0, X=rbind(a=1, b=1:3))  # use some names
xx <- data.frame(I=rep(0,2))
cbind(xx, X=rbind(a=1, b=1:3))   # named differently

cbind(0, matrix(1, nrow=0, ncol=4)) # Warning (making sense)
dim(cbind(0, matrix(1, nrow=2, ncol=0))) # 2 x 1
```

char.expand *Expand a String with Respect to a Target Table*

Description

Seeks a unique match of its first argument among the elements of its second. If successful, it returns this element; otherwise, it performs an action specified by the third argument.

Usage

```
char.expand(input, target, nomatch = stop("no match"))
```

Arguments

`input`	a character string to be expanded.
`target`	a character vector with the values to be matched against.
`nomatch`	an R expression to be evaluated in case expansion was not possible.

Details

This function is particularly useful when abbreviations are allowed in function arguments, and need to be uniquely expanded with respect to a target table of possible values.

See Also

`charmatch` and `pmatch` for performing partial string matching.

Examples

```
locPars <- c("mean", "median", "mode")
char.expand("me", locPars, warning("Could not expand!"))
char.expand("mo", locPars)
```

character *Character Vectors*

Description

Create or test for objects of type "character".

Usage

```
character(length = 0)
as.character(x, ...)
is.character(x)
```

Arguments

length	desired length.
x	object to be coerced or tested.
...	further arguments passed to or from other methods.

Details

as.character and is.character are generic: you can write methods to handle specific classes of objects, see InternalMethods.

Value

character creates a character vector of the specified length. The elements of the vector are all equal to "".

as.character attempts to coerce its argument to character type; like as.vector it strips attributes including names.

is.character returns TRUE or FALSE depending on whether its argument is of character type or not.

Note

as.character truncates components of language objects to 500 characters (was about 70 before 1.3.1).

References

Becker, R. A., Chambers, J. M. and Wilks, A. R. (1988) *The New S Language.* Wadsworth & Brooks/Cole.

See Also

paste, **substr** and **strsplit** for character concatenation and splitting, **chartr** for character translation and casefolding (e.g., upper to lower case) and **sub**, **grep** etc for string matching and substitutions. Note that `help.search(keyword = "character")` gives even more links. **deparse**, which is normally preferable to **as.character** for language objects.

Examples

```
form <- y ~ a + b + c
as.character(form)  ## length 3
deparse(form)       ## like the input
```

charmatch *Partial String Matching*

Description

charmatch seeks matches for the elements of its first argument among those of its second.

Usage

```
charmatch(x, table, nomatch = NA)
```

Arguments

x	the values to be matched.
table	the values to be matched against.
nomatch	the value returned at non-matching positions.

Details

Exact matches are preferred to partial matches (those where the value to be matched has an exact match to the initial part of the target, but the target is longer).

If there is a single exact match or no exact match and a unique partial match then the index of the matching value is returned; if multiple exact or multiple partial matches are found then 0 is returned and if no match is found then NA is returned.

Author(s)

This function is based on a C function written by Terry Therneau.

See Also

pmatch, match.

grep or regexpr for more general (regexp) matching of strings.

Examples

```
charmatch("", "")                             # returns 1
charmatch("m",   c("mean", "median", "mode")) # returns 0
charmatch("med", c("mean", "median", "mode")) # returns 2
```

`chartr` *Character Translation and Casefolding*

Description

Translate characters in character vectors, in particular from upper to lower case or vice versa.

Usage

```
chartr(old, new, x)
tolower(x)
toupper(x)
casefold(x, upper = FALSE)
```

Arguments

`x`	a character vector.
`old`	a character string specifying the characters to be translated.
`new`	a character string specifying the translations.
`upper`	logical: translate to upper or lower case?.

Details

`chartr` translates each character in `x` that is specified in `old` to the corresponding character specified in `new`. Ranges are supported in the specifications, but character classes and repeated characters are not. If `old` contains more characters than new, an error is signaled; if it contains fewer characters, the extra characters at the end of `new` are ignored.

`tolower` and `toupper` convert upper-case characters in a character vector to lower-case, or vice versa. Non-alphabetic characters are left unchanged.

`casefold` is a wrapper for `tolower` and `toupper` provided for compatibility with S-PLUS.

See Also

`sub` and `gsub` for other substitutions in strings.

Examples

```
x <- "MiXeD cAsE 123"
chartr("iXs", "why", x)
chartr("a-cX", "D-Fw", x)
tolower(x)
toupper(x)
```

check.options *Set Options with Consistency Checks*

Description

Utility function for setting options with some consistency checks. The `attributes` of the new settings in `new` are checked for consistency with the *model* (often default) list in `name.opt`.

Usage

```
check.options(new, name.opt, reset = FALSE,
              assign.opt = FALSE,
              envir = .GlobalEnv,
              check.attributes = c("mode", "length"),
              override.check = FALSE)
```

Arguments

`new`	a *named* list
`name.opt`	character with the name of R object containing the "model" (default) list.
`reset`	logical; if `TRUE`, reset the options from `name.opt`. If there is more than one R object with name `name.opt`, remove the first one in the `search()` path.
`assign.opt`	logical; if `TRUE`, assign the ...
`envir`	the `environment` used for `get` and `assign`.
`check.attributes`	character containing the attributes which `check.options` should check.
`override.check`	logical vector of length `length(new)` (or 1 which entails recycling). For those `new[i]` where `override.check[i] == TRUE`, the checks are overriden and the changes made anyway.

Value

A list of components with the same names as the one called `name.opt`. The values of the components are changed from the `new` list, as long as these pass the checks (when these are not overridden according to `override.check`).

Author(s)

Martin Maechler

See Also

`ps.options` which uses `check.options`.

Examples

```
L1 <- list(a=1:3, b=pi, ch="CH")
str(L2 <- check.options(list(a=0:2), name.opt = "L1"))
str(check.options(NULL, reset = TRUE, name.opt = "L1"))
```

`citation` *Citing R in Publications*

Description

How to cite R in publications.

Usage

```
citation()
```

Details

Execute function `citation()` for information on how to cite R in publications.

class *Object Classes*

Description

R possesses a simple generic function mechanism which can be used for an object-oriented style of programming. Method dispatch takes place based on the class of the first argument to the generic function.

Usage

```
class(x)
class(x) <- value
unclass(x)
inherits(x, what, which = FALSE)

oldClass(x)
oldClass(x) <- value
```

Arguments

`x`	a R object
`what, value`	a character vector naming classes.
`which`	logical affecting return value: see Details.

Details

Many R objects have a `class` attribute, a character vector giving the names of the classes which the object "inherits" from. If the object does not have a class attribute, it has an implicit class, `"matrix"`, `"array"` or the result of `mode(x)`. (Functions `oldClass` and `oldClass<-` get and set the attribute, which can also be done directly.)

When a generic function `fun` is applied to an object with class attribute `c("first", "second")`, the system searches for a function called `fun.first` and, if it finds it, applies it to the object. If no such function is found, a function called `fun.second` is tried. If no class name produces a suitable function, the function `fun.default` is used (if it exists). If there is no class attribute, the implicit class is tried, then the default method.

The function `class` prints the vector of names of classes an object inherits from. Correspondingly, `class<-` sets the classes an object inherits from.

`unclass` returns (a copy of) its argument with its class attribute removed.

`inherits` indicates whether its first argument inherits from any of the classes specified in the `what` argument. If `which` is `TRUE` then an integer vector of the same length as `what` is returned. Each element indicates the position in the `class(x)` matched by the element of `what`; zero indicates no match. If `which` is `FALSE` then `TRUE` is returned by `inherits` if any of the names in `what` match with any `class`.

Formal classes

An additional mechanism of *formal* classes has been available in packages **methods** since R 1.4.0, and as from R 1.7.0 this is attached by default. For objects which have a formal class, its name is returned by `class` as a character vector of length one.

The replacement version of the function sets the class to the value provided. For classes that have a formal definition, directly replacing the class this way is strongly deprecated. The expression `as(object, value)` is the way to coerce an object to a particular class.

Note

Functions `oldClass` and `oldClass<-` behave in the same way as functions of those names in S-PLUS 5/6, *but* in R `UseMethod` dispatches on the class as returned by `class` rather than `oldClass`.

See Also

`UseMethod`, `NextMethod`.

Examples

```
x <- 10
inherits(x, "a") # FALSE
class(x)<-c("a", "b")
inherits(x,"a") # TRUE
inherits(x, "a", TRUE) # 1
inherits(x, c("a", "b", "c"), TRUE) # 1 2 0
```

`close.socket` *Close a Socket*

Description

Closes the socket and frees the space in the file descriptor table. The port may not be freed immediately.

Usage

```
close.socket(socket, ...)
```

Arguments

`socket`	A `socket` object
`...`	further arguments passed to or from other methods.

Value

logical indicating success or failure

Author(s)

Thomas Lumley

See Also

`make.socket`, `read.socket`

`codes-deprecated` *Factor Codes*

Description

This (generic) function returns a numeric coding of a factor. It can also be used to assign to a factor using the coded form.

It is now `Deprecated`.

Usage

```
codes(x, ...)
codes(x, ...) <- value
```

Arguments

`x`	an object from which to extract or set the codes.
`...`	further arguments passed to or from other methods.
`value`	replacement value.

Value

For an ordered factor, it returns the internal coding (1 for the lowest group, 2 for the second lowest, etc.).

For an unordered factor, an alphabetical ordering of the levels is assumed, i.e., the level that is coded 1 is the one whose name is sorted first according to the prevailing collating sequence. **Warning:** the sort order may well depend on the locale, and should not be assumed to be ASCII.

Note

Normally `codes` is not the appropriate function to use with an unordered factor. Use `unclass` or `as.numeric` to extract the codes used in the internal representation of the factor, as these do not assume that the codes are sorted.

The behaviour for unordered factors is dubious, but compatible with S version 3. To get the internal coding of a factor, use `as.integer`. Note in particular that the codes may not be the same in different language locales because of collating differences.

See Also

`factor`, `levels`, `nlevels`.

Examples

```
codes(rep(factor(c(20,10)),3))

x <- gl(3,5)
codes(x)[3] <- 2
x

data(esoph)
( ag <- esoph$alcgp[12:1] )
codes(ag)

codes(factor(1:10)) # BEWARE!
```

col *Column Indexes*

Description

Returns a matrix of integers indicating their column number in the matrix.

Usage

```
col(x, as.factor=FALSE)
```

Arguments

`x`	a matrix.
`as.factor`	a logical value indicating whether the value should be returned as a factor rather than as numeric.

Value

An integer matrix with the same dimensions as `x` and whose `ij`-th element is equal to `j`.

References

Becker, R. A., Chambers, J. M. and Wilks, A. R. (1988) *The New S Language.* Wadsworth & Brooks/Cole.

See Also

`row` to get rows.

Examples

```
# extract an off-diagonal of a matrix
ma <- matrix(1:12, 3, 4)
ma[row(ma) == col(ma) + 1]

# create an identity 5-by-5 matrix
x <- matrix(0, nr = 5, nc = 5)
x[row(x) == col(x)] <- 1
```

commandArgs *Extract Command Line Arguments*

Description

Provides access to a copy of the command line arguments supplied when this R session was invoked.

Usage

```
commandArgs()
```

Details

These arguments are captured before the standard R command line processing takes place. This means that they are the unmodified values. If it were useful, we could provide support an argument which indicated whether we want the unprocessed or processed values.

This is especially useful with the `--args` command-line flag to R, as all of the command line after than flag is skipped.

Value

A character vector containing the name of the executable and the user-supplied command line arguments. The first element is the name of the executable by which R was invoked. As far as I am aware, the exact form of this element is platform dependent. It may be the fully qualified name, or simply the last component (or basename) of the application.

See Also

`BATCH`

Examples

```
commandArgs()
## Spawn a copy of this application as it was invoked.
## system(paste(commandArgs(), collapse=" "))
```

comment *Query or Set a 'Comment' Attribute*

Description

These functions set and query a *comment* attribute for any R objects. This is typically useful for `data.frames` or model fits.

Contrary to other `attributes`, the `comment` is not printed (by `print` or `print.default`).

Usage

```
comment(x)
comment(x) <- value
```

Arguments

`x`	any R object
`value`	a character vector

See Also

`attributes` and `attr` for "normal" attributes.

Examples

```
x <- matrix(1:12, 3,4)
comment(x) <- c("This is the data from experiment #0234",
                "Jun 5, 1998")
x
comment(x)
```

`Comparison` *Relational Operators*

Description

Binary operators which allow the comparison of values in vectors.

Usage

```
x < y
x > y
x <= y
x >= y
x == y
x != y
```

Details

Comparison of strings in character vectors is lexicographic within the strings using the collating sequence of the locale in use: see `locales`. The collating sequence of locales such as '`en_US`' is normally different from '`C`' (which should use ASCII) and can be surprising.

Value

A vector of logicals indicating the result of the element by element comparison. The elements of shorter vectors are recycled as necessary.

Objects such as arrays or time-series can be compared this way provided they are conformable.

Note

Don't use == and != for tests, such as in `if` expressions, where you must get a single `TRUE` or `FALSE`. Unless you are absolutely sure that nothing unusual can happen, you should use the `identical` function instead.

For numerical values, remember == and != do not allow for the finite representation of fractions, nor for rounding error. Using `all.equal` with `identical` is almost always preferable. See the examples.

References

Becker, R. A., Chambers, J. M. and Wilks, A. R. (1988) *The New S Language*. Wadsworth & Brooks/Cole.

See Also

`Syntax` for operator precedence.

Examples

```
x <- rnorm(20)
x < 1
x[x > 0]

x1 <- 0.5 - 0.3
x2 <- 0.3 - 0.1
x1 == x2                            # FALSE on most machines
identical(all.equal(x1, x2), TRUE) # TRUE everywhere
```

COMPILE *Compile Files for Use with R*

Description

Compile given source files so that they can subsequently be collected into a shared library using `R CMD SHLIB` and be loaded into R using `dyn.load()`.

Usage

```
R CMD COMPILE [options] srcfiles
```

Arguments

`srcfiles`	A list of the names of source files to be compiled. Currently, C, C++ and FORTRAN are supported; the corresponding files should have the extensions '`.c`', '`.cc`' (or '`.cpp`' or '`.C`'), and '`.f`', respectively.
`options`	A list of compile-relevant settings, such as special values for `CFLAGS` or `FFLAGS`, or for obtaining information about usage and version of the utility.

Details

Note that Ratfor is not supported. If you have Ratfor source code, you need to convert it to FORTRAN. On many Solaris systems mixing Ratfor and FORTRAN code will work.

See Also

`SHLIB`, `dyn.load`

complete.cases *Find Complete Cases*

Description

Return a logical vector indicating which cases are complete, i.e., have no missing values.

Usage

```
complete.cases(...)
```

Arguments

`...`	a sequence of vectors, matrices and data frames.

Value

A logical vector specifying which observations/rows have no missing values across the entire sequence.

See Also

`is.na`, `na.omit`, `na.fail`.

Examples

```
data(airquality)
x <- airquality[, -1] # x is a regression design matrix
y <- airquality[,  1] # y is the corresponding response

stopifnot(complete.cases(y) != is.na(y))
ok <- complete.cases(x,y)
sum(!ok) # how many are not "ok" ?
x <- x[ok,]
y <- y[ok]
```

`complex` *Complex Vectors*

Description

Basic functions which support complex arithmetic in R.

Usage

```
complex(length.out = 0,
        real = numeric(), imaginary = numeric(),
        modulus = 1, argument = 0)
as.complex(x, ...)
is.complex(x)

Re(x)
Im(x)
Mod(x)
Arg(x)
Conj(x)
```

Arguments

`length.out`	numeric. Desired length of the output vector, inputs being recycled as needed.
`real`	numeric vector.
`imaginary`	numeric vector.
`modulus`	numeric vector.
`argument`	numeric vector.
`x`	an object, probably of mode `complex`.
`...`	further arguments passed to or from other methods.

Details

Complex vectors can be created with `complex`. The vector can be specified either by giving its length, its real and imaginary parts, or modulus and argument. (Giving just the length generates a vector of complex zeroes.)

`as.complex` attempts to coerce its argument to be of complex type: like `as.vector` it strips attributes including names.

Note that `is.complex` and `is.numeric` are never both `TRUE`.

The functions `Re`, `Im`, `Mod`, `Arg` and `Conj` have their usual interpretation as returning the real part, imaginary part, modulus, argument and complex conjugate for complex values. Modulus and argument are also called the *polar coordinates*. If $z = x + iy$ with real x and y, $\texttt{Mod}(z) = \sqrt{x^2 + y^2}$, and for $\phi = Arg(z)$, $x = \cos(\phi)$ and $y = \sin(\phi)$.

In addition, the elementary trigonometric, logarithmic and exponential functions are available for complex values.

`is.complex` is generic: you can write methods to handle specific classes of objects, see InternalMethods.

References

Becker, R. A., Chambers, J. M. and Wilks, A. R. (1988) *The New S Language*. Wadsworth & Brooks/Cole.

Examples

```
0i ^ (-3:3)

matrix(1i^ (-6:5), nr=4) # all columns are the same
0 ^ 1i # a complex NaN

## create a complex normal vector
z <- complex(real = rnorm(100), imag = rnorm(100))
## or also (less efficiently):
z2 <- 1:2 + 1i*(8:9)

## The Arg(.) is an angle:
zz <- (rep(1:4,len=9) + 1i*(9:1))/10
zz.shift <- complex(modulus = Mod(zz),
                    argument= Arg(zz) + pi)
plot(zz, xlim=c(-1,1), ylim=c(-1,1), col="red", asp = 1,
  main = expression(paste("Rotation by "," ", pi == 180^o)))
abline(h=0,v=0, col="blue", lty=3)
points(zz.shift, col="orange")
```

conditions *Condition Handling and Recovery*

Description

These functions provide a mechanism for handling unusual conditions, including errors and warnings.

Usage

```
tryCatch(expr, ..., finally)
withCallingHandlers(expr, ...)

signalCondition(cond)

simpleCondition(message, call = NULL)
simpleError    (message, call = NULL)
simpleWarning  (message, call = NULL)

## S3 method for class 'condition':
as.character(x, ...)
## S3 method for class 'error':
as.character(x, ...)
## S3 method for class 'condition':
print(x, ...)
## S3 method for class 'restart':
print(x, ...)

conditionCall(c)
## S3 method for class 'condition':
conditionCall(c)
conditionMessage(c)
## S3 method for class 'condition':
conditionMessage(c)

withRestarts(expr, ...)

computeRestarts(cond = NULL)
findRestart(name, cond = NULL)
invokeRestart(r, ...)
invokeRestartInteractively(r)
```

```
isRestart(x)
restartDescription(r)
restartFormals(r)

.signalSimpleWarning(msg, call)
.handleSimpleError(h, msg, call)
```

Arguments

`c`	a condition object.
`call`	call expression.
`cond`	a condition object.
`expr`	expression to be evaluated.
`finally`	expression to be evaluated before returning or exiting.
`h`	function.
`message`	character string.
`msg`	character string.
`name`	character string naming a restart.
`r`	restart object.
`x`	object.
`...`	additional arguments; see details below.

Details

The condition system provides a mechanism for signaling and handling unusual conditions, including errors and warnings. Conditions are represented as objects that contain information about the condition that occurred, such as a message and the call in which the condition occurred. Currently conditions are S3-style objects, though this may eventually change.

Conditions are objects inheriting from the abstract class `condition`. Errors and warnings are objects inheriting from the abstract subclasses `error` and `warning`. The class `simpleError` is the class used by `stop` and all internal error signals. Similarly, `simpleWarning` is used by `warning`. The constructors by the same names take a string describing the condition as argument and an optional call. The functions `conditionMessage` and `conditionCall` are generic functions that return the message and call of a condition.

Conditions are signaled by `signalCondition`. In addition, the `stop` and `warning` functions have been modified to also accept condition arguments.

The function `tryCatch` evaluates its expression argument in a context where the handlers provided in the `...` argument are available. The `finally` expression is then evaluated in the context in which `tryCatch` was called; that is, the handlers supplied to the current `tryCatch` call are not active when the `finally` expression is evaluated.

Handlers provided in the `...` argument to `tryCatch` are established for the duration of the evaluation of `expr`. If no condition is signaled when evaluating `expr` then `tryCatch` returns the value of the expression.

If a condition is signaled while evaluating `expr` then established handlers are checked, starting with the most recently established ones, for one matching the class of the condition. When several handlers are supplied in a single `tryCatch` then the first one is considered more recent than the second. If a handler is found then control is transferred to the `tryCatch` call that established the handler, the handler found and all more recent handlers are disestablished, the handler is called with the condition as its argument, and the result returned by the handler is returned as the value of the `tryCatch` call.

Calling handlers are established by `withCallingHandlers`. If a condition is signaled and the applicable handler is a calling handler, then the handler is called by `signalCondition` in the context where the condition was signaled but with the available handlers restricted to those below the handler called in the handler stack. If the handler returns, then the next handler is tried; once the last handler has been tried, `signalCondition` returns `NULL`.

User interrupts signal a condition of class `interrupt` that inherits directly from class `condition` before executing the default interrupt action.

Restarts are used for establishing recovery protocols. They can be established using `withRestarts`. One pre-established restart is an `abort` restart that represents a jump to top level.

`findRestart` and `computeRestarts` find the available restarts. `findRestart` returns the most recently established restart of the specified name. `computeRestarts` returns a list of all restarts. Both can be given a condition argument and will then ignore restarts that do not apply to the condition.

`invokeRestart` transfers control to the point where the specified restart was established and calls the restart's handler with the arguments, if any, given as additional arguments to `invokeRestart`. The restart argument to `invokeRestart` can be a character string, in which case `findRestart` is used to find the restart.

New restarts for `withRestarts` can be specified in several ways. The simplest is in `name=function` form where the function is the handler to call when the restart is invoked. Another simple variant is as `name=string` where the string is stored in the `description` field of the restart object returned by `findRestart`; in this case the handler ignores its arguments and returns `NULL`. The most flexible form of a restart specification is as a list that can include several fields, including `hander`, `description`, and `test`. The `test` field should contain a function of one argument, a condition, that returns `TRUE` if the restart applies to the condition and `FALSE` if it does not; the default function returns `TRUE` for all conditions.

One additional field that can be specified for a restart is `interactive`. This should be a function of no arguments that returns a list of arguments to pass to the restart handler. The list could be obtained by interacting with the user if necessary. The function `invokeRestartInteractively` calls this function to obtain the arguments to use when invoking the restart. The default `interactive` method queries the user for values for the formal arguments of the handler function.

`.signalSimpleWarning` and `.handleSimpleError` are used internally and should not be called directly.

References

The `tryCatch` mechanism is similar to Java error handling. Calling handlers are based on Common Lisp and Dylan. Restarts are based on the Common Lisp restart mechanism.

See Also

`stop` and `warning` signal conditions, and `try` is essentially a simplified version of `tryCatch`.

Examples

```
tryCatch(1, finally=print("Hello"))
e <- simpleError("test error")
stop(e)
tryCatch(stop(e), finally=print("Hello"))
tryCatch(stop("fred"), finally=print("Hello"))
```

```
tryCatch(stop(e), error = function(e) e,
         finally=print("Hello"))
tryCatch(stop("fred"),  error = function(e) e,
         finally=print("Hello"))
withCallingHandlers({ warning("A"); 1+2 },
                    warning = function(w) {})
{ try(invokeRestart("tryRestart")); 1}
{ withRestarts(stop("A"), abort = function() {}); 1}
withRestarts(invokeRestart("foo", 1, 2),
             foo = function(x, y) {x + y})
```

`conflicts` *Search for Masked Objects on the Search Path*

Description

`conflicts` reports on objects that exist with the same name in two or more places on the `search` path, usually because an object in the user's workspace or a package is masking a system object of the same name. This helps discover unintentional masking.

Usage

```
conflicts(where=search(), detail=FALSE)
```

Arguments

`where`	A subset of the search path, by default the whole search path.
`detail`	If `TRUE`, give the masked or masking functions for all members of the search path.

Value

If `detail=FALSE`, a character vector of masked objects. If `detail=TRUE`, a list of character vectors giving the masked or masking objects in that member of the search path. Empty vectors are omitted.

Examples

```
lm <- 1:3
conflicts(, TRUE)
## gives something like
# $.GlobalEnv
# [1] "lm"
#
# $package:base
# [1] "lm"

## Remove things from your "workspace" that mask others:
remove(list = conflicts(detail=TRUE)$.GlobalEnv)
```

`connections` *Functions to Manipulate Connections*

Description

Functions to create, open and close connections.

Usage

```
file(description = "", open = "", blocking = TRUE,
     encoding = getOption("encoding"))
pipe(description, open = "",
     encoding = getOption("encoding"))
fifo(description = "", open = "", blocking = FALSE,
     encoding = getOption("encoding"))
gzfile(description, open = "",
       encoding = getOption("encoding"),
       compression = 6)
unz(description, filename, open = "",
    encoding = getOption("encoding"))
bzfile(description, open = "",
       encoding = getOption("encoding"))
url(description, open = "", blocking = TRUE,
    encoding = getOption("encoding"))
socketConnection(host = "localhost", port, server = FALSE,
                 blocking = FALSE, open = "a+",
                 encoding = getOption("encoding"))

open(con, ...)
## S3 method for class 'connection':
open(con, open = "r", blocking = TRUE, ...)
close(con, ...)
## S3 method for class 'connection':
close(con, type = "rw", ...)

flush(con)

isOpen(con, rw = "")
isIncomplete(con)
```

Arguments

`description`	character. A description of the connection. For `file` and `pipe` this is a path to the file to be opened. For `url` it is a complete URL, including schemes (`http://`, `ftp://` or `file://`). `file` also accepts complete URLs.
`filename`	a filename within a zip file.
`con`	a connection.
`host`	character. Host name for port.
`port`	integer. The TCP port number.
`server`	logical. Should the socket be a client or a server?
`open`	character. A description of how to open the connection (if at all). See Details for possible values.
`blocking`	logical. See 'Blocking' section below.
`encoding`	An integer vector of length 256.
`compression`	integer in 0–9. The amount of compression to be applied when writing, from none to maximal. The default is a good space/time compromise.
`type`	character. Currently ignored.
`rw`	character. Empty or `"read"` or `"write"`, partial matches allowed.
`...`	arguments passed to or from other methods.

Details

The first eight functions create connections. By default the connection is not opened (except for `socketConnection`), but may be opened by setting a non-empty value of argument `open`.

`gzfile` applies to files compressed by '`gzip`', and `bzfile` to those compressed by '`bzip2`': such connections can only be binary.

`unz` reads (only) single files within zip files, in binary mode. The description is the full path, with '`.zip`' extension if required.

All platforms support (gz)file connections and `url("file://")` connections. The other types may be partially implemented or not implemented at all. (They do work on most Unix platforms, and all but `fifo` on Windows.)

Proxies can be specified for `url` connections: see `download.file`.

`open`, `close` and `seek` are generic functions: the following applies to the methods relevant to connections.

`open` opens a connection. In general, functions using connections will open them if they are not open, but then close them again, so to leave a connection open call `open` explicitly.

Possible values for the mode `open` to open a connection are

`"r"` or `"rt"` Open for reading in text mode.

`"w"` or `"wt"` Open for writing in text mode.

`"a"` or `"at"` Open for appending in text mode.

`"rb"` Open for reading in binary mode.

`"wb"` Open for writing in binary mode.

`"ab"` Open for appending in binary mode.

`"r+"`, `"r+b"` Open for reading and writing.

`"w+"`, `"w+b"` Open for reading and writing, truncating file initially.

`"a+"`, `"a+b"` Open for reading and appending.

Not all modes are applicable to all connections: for example URLs can only be opened for reading. Only file and socket connections can be opened for reading and writing/appending. For many connections there is little or no difference between text and binary modes, but there is for file-like connections on Windows, and `pushBack` is text-oriented and is only allowed on connections open for reading in text mode.

`close` closes and destroys a connection.

`flush` flushes the output stream of a connection open for write/append (where implemented).

If for a `file` connection the description is `""`, the file is immediately opened in `"w+"` mode and unlinked from the file system. This provides a temporary file to write to and then read from.

The encoding vector is used to map the input from a file or pipe to the platform's native character set. Supplied examples are `native.enc` as well as `MacRoman`, `WinAnsi` and `ISOLatin1`, whose actual encoding is platform-dependent. Missing characters are mapped to a space in these encodings.

Value

`file`, `pipe`, `fifo`, `url`, `gzfile` and `socketConnection` return a connection object which inherits from class `"connection"` and has a first more specific class.

`isOpen` returns a logical value, whether the connection is currently open.

`isIncomplete` returns a logical value, whether last read attempt was blocked, or for an output text connection whether there is unflushed output.

Blocking

The default condition for all but fifo and socket connections is to be in blocking mode. In that mode, functions do not return to the R evaluator until they are complete. In non-blocking mode, operations return as soon as possible, so on input they will return with whatever input is available (possibly none) and for output they will return whether or not the write succeeded.

The function `readLines` behaves differently in respect of incomplete last lines in the two modes: see its help page.

Even when a connection is in blocking mode, attempts are made to ensure that it does not block the event loop and hence the operation of GUI parts of R. These do not always succeed, and the whole process will be blocked during a DNS lookup on Unix, for example.

Most blocking operations on URLs and sockets are subject to the time-out set by `options("timeout")`. Note that this is a timeout for no response at all, not for the whole operation.

Fifos

Fifos default to non-blocking. That follows Svr4 and it probably most natural, but it does have some implications. In particular, opening a non-blocking fifo connection for writing (only) will fail unless some other process is reading on the fifo.

Opening a fifo for both reading and writing (in any mode: one can only append to fifos) connects both sides of the fifo to the R process, and provides a similar facility to `file()`.

Note

R's connections are modelled on those in S version 4 (see Chambers, 1998). However R goes well beyond the Svr4 model, for example in output text connections and URL, gzfile, bzfile and socket connections.

The default mode in R is `"r"` except for socket connections. This differs from Svr4, where it is the equivalent of `"r+"`, known as `"*"`.

On platforms where `vsnprintf` does not return the needed length of output (e.g., Windows) there is a 100,000 character output limit on the length of line for `fifo`, `gzfile` and `bzfile` connections: longer lines will be truncated with a warning.

References

Chambers, J. M. (1998) *Programming with Data. A Guide to the S Language.* Springer.

See Also

`textConnection`, `seek`, `readLines`, `readBin`, `writeLines`, `writeBin`, `showConnections`, `pushBack`.

`capabilities` to see if `gzfile`, `url`, `fifo` and `socketConnection` are supported by this build of R.

Examples

```
zz <- file("ex.data", "w") # open an output file connection
cat("TITLE extra line", "2 3 5 7", "", "11 13 17",
    file = zz, sep = "\n")
cat("One more line\n", file = zz)
close(zz)
readLines("ex.data")
unlink("ex.data")

zz <- gzfile("ex.gz", "w")  # compressed file
cat("TITLE extra line", "2 3 5 7", "", "11 13 17",
    file = zz, sep = "\n")
close(zz)
readLines(gzfile("ex.gz"))
unlink("ex.gz")

if(capabilities("bzip2")) {
  zz <- bzfile("ex.bz2", "w")  # bzip2-ed file
  cat("TITLE extra line", "2 3 5 7", "", "11 13 17",
      file = zz, sep = "\n")
  close(zz)
  print(readLines(bzfile("ex.bz2")))
  unlink("ex.bz2")
}

## An example of a file open for reading and writing
Tfile <- file("test1", "w+")
```

```
c(isOpen(Tfile, "r"), isOpen(Tfile, "w")) # both TRUE
cat("abc\ndef\n", file=Tfile)
readLines(Tfile)
seek(Tfile, 0, rw="r") # reset to beginning
readLines(Tfile)
cat("ghi\n", file=Tfile)
readLines(Tfile)
close(Tfile)
unlink("test1")

## We can do the same thing with an anonymous file.
Tfile <- file()
cat("abc\ndef\n", file=Tfile)
readLines(Tfile)
close(Tfile)

if(capabilities("fifo")) {
  zz <- fifo("foo", "w+")
  writeLines("abc", zz)
  print(readLines(zz))
  close(zz)
  unlink("foo")
}

## Unix examples of use of pipes

# read listing of current directory
readLines(pipe("ls -1"))

# remove trailing commas. Suppose
% cat data2
450, 390, 467, 654,  30, 542, 334, 432, 421,
357, 497, 493, 550, 549, 467, 575, 578, 342,
446, 547, 534, 495, 979, 479
# Then read this by
scan(pipe("sed -e s/,$// data2"), sep=",")

# convert decimal point to comma in output
# both R strings and (probably) the shell need \ doubled
zz <- pipe(paste("sed s/\\\\./,/ >", "outfile"), "w")
cat(format(round(rnorm(100), 4)), sep = "\n", file = zz)
close(zz)
file.show("outfile", delete.file=TRUE)
```

```
## example for Unix machine running a finger daemon
con <- socketConnection(port = 79, blocking = TRUE)
writeLines(paste(system("whoami", intern=TRUE), "\r",
           sep=""), con)
gsub(" *$", "", readLines(con))
close(con)

## two R processes communicating via non-blocking sockets
# R process 1
con1 <- socketConnection(port = 6011, server=TRUE)
writeLines(LETTERS, con1)
close(con1)

# R process 2
con2 <- socketConnection(Sys.info()["nodename"],
                         port = 6011)
# as non-blocking, may need to loop for input
readLines(con2)
while(isIncomplete(con2)) {Sys.sleep(1); readLines(con2)}
close(con2)
```

Constants *Built-in Constants*

Description

Constants built into R.

Usage

```
LETTERS
letters
month.abb
month.name
pi
```

Details

R has a limited number of built-in constants (there is also a rather larger library of data sets which can be loaded with the function `data`).

The following constants are available:

- `LETTERS`: the 26 upper-case letters of the Roman alphabet;
- `letters`: the 26 lower-case letters of the Roman alphabet;
- `month.abb`: the three-letter abbreviations for the English month names;
- `month.name`: the English names for the months of the year;
- `pi`: the ratio of the circumference of a circle to its diameter.

References

Becker, R. A., Chambers, J. M. and Wilks, A. R. (1988) *The New S Language*. Wadsworth & Brooks/Cole.

See Also

`data`.

Examples

```
# John Machin (1705) computed 100 decimals of pi :
pi - 4*(4*atan(1/5) - atan(1/239))
```

`contributors` *R Project Contributors*

Description

The R Who-is-who, describing who made significant contributions to the development of R.

Usage

```
contributors()
```

`Control` *Control Flow*

Description

These are the basic control-flow constructs of the R language. They function in much the same way as control statements in any algol-like language.

Usage

```
if(cond) expr
if(cond) cons.expr  else  alt.expr
for(var in seq) expr
while(cond) expr
repeat expr
break
next
```

Details

Note that `expr` and `cons.expr`, etc, in the Usage section above means an *expression* in a formal sense. This is either a simple expression or a so called *compound expression*, usually of the form { `expr1 ; expr2` }.

Note that it is a common mistake to forget putting braces ({ `..` }) around your statements, e.g., after `if(..)` or `for(....)`. For that reason, one (somewhat extreme) attitude of defensive programming uses braces always, e.g., for `if` clauses.

The index `seq` in a `for` loop is evaluated at the start of the loop; changing it subsequently does not affect the loop.

References

Becker, R. A., Chambers, J. M. and Wilks, A. R. (1988) *The New S Language.* Wadsworth & Brooks/Cole.

See Also

`Syntax` for the basic R syntax and operators, `Paren` for parentheses and braces; further, `ifelse`, `switch`.

Examples

```
for(i in 1:5) print(1:i)
for(n in c(2,5,10,20,50)) {
   x <- rnorm(n)
   cat(n,":", sum(x^2),"\n")
}
```

`copyright` *Copyrights of Files Used to Build R*

Description

R is released under the 'GNU General Public License': see `license` for details. The license describes your right to use R. Some of the software used has conditions that the copyright must be explicitly stated: see the Details section. We are grateful to these people and other contributors (see `contributors`) for the ability to use their work.

Details

The file '`$R_HOME/COPYRIGHTS`' lists the copyrights in full detail.

cor *Correlation, Variance and Covariance (Matrices)*

Description

`var`, `cov` and `cor` compute the variance of `x` and the covariance or correlation of `x` and `y` if these are vectors. If `x` and `y` are matrices then the covariances (or correlations) between the columns of `x` and the columns of `y` are computed.

`cov2cor` scales a covariance matrix into the corresponding correlation matrix *efficiently*.

Usage

```
var(x, y = NULL, na.rm = FALSE, use)
cov(x, y = NULL, use = "all.obs",
    method = c("pearson", "kendall", "spearman"))
cor(x, y = NULL, use = "all.obs",
    method = c("pearson", "kendall", "spearman"))
cov2cor(V)
```

Arguments

`x`	a numeric vector, matrix or data frame.
`y`	`NULL` (default) or a vector, matrix or data frame with compatible dimensions to `x`. The default is equivalent to `y = x` (but more efficient).
`na.rm`	logical. Should missing values be removed?
`use`	an optional character string giving a method for computing covariances in the presence of missing values. This must be (an abbreviation of) one of the strings `"all.obs"`, `"complete.obs"` or `"pairwise.complete.obs"`.
`method`	a character string indicating which correlation coefficient (or covariance) is to be computed. One of `"pearson"` (default), `"kendall"`, or `"spearman"`, can be abbreviated.
`V`	symmetric numeric matrix, usually positive definite such as a covariance matrix.

Details

For `cov` and `cor` one must *either* give a matrix or data frame for `x` *or* give both `x` and `y`.

`var` is just another interface to `cov`, where `na.rm` is used to determine the default for `use` when that is unspecified. If `na.rm` is `TRUE` then the complete observations (rows) are used (`use = "complete"`) to compute the variance. Otherwise (`use = "all"`), `var` will give an error if there are missing values.

If `use` is `"all.obs"`, then the presence of missing observations will produce an error. If `use` is `"complete.obs"` then missing values are handled by casewise deletion. Finally, if `use` has the value `"pairwise.complete.obs"` then the correlation between each pair of variables is computed using all complete pairs of observations on those variables. This can result in covariance or correlation matrices which are not positive semidefinite.

The denominator $n - 1$ is used which gives an unbiased estimator of the (co)variance for i.i.d. observations. These functions return `NA` when there is only one observation (whereas S-plus has been returning `NaN`), and fail if `x` has length zero.

For `cor()`, if `method` is `"kendall"` or `"spearman"`, Kendall's τ or Spearman's ρ statistic is used to estimate a rank-based measure of association. These are more robust and have be recommended if the data do not necessarily come from a bivariate normal distribution.
For `cov()`, a non-Pearson method is unusual but available for the sake of completeness.

Scaling a covariance matrix into a correlation one can be achieved in many ways, mathematically most appealing by multiplication with a diagonal matrix from left and right, or more efficiently by using `sweep(.., FUN = "/")` twice. The `cov2cor` function is even a bit more efficient, and provided mostly for didactical reasons.

Value

For `r <- cor(*, use = "all.obs")`, it is now guaranteed that `all(r <= 1)`.

References

Becker, R. A., Chambers, J. M. and Wilks, A. R. (1988) *The New S Language.* Wadsworth & Brooks/Cole.

See Also

`cor.test` (package **ctest**) for confidence intervals (and tests).
`cov.wt` for *weighted* covariance computation, `sd` for standard deviation (vectors).

Examples

```
var(1:10) # 9.166667

var(1:5,1:5) # 2.5

## Two simple vectors
cor(1:10,2:11) # == 1

## Correlation Matrix of Multivariate sample:
data(longley)
(Cl <- cor(longley))
## Graphical Correlation Matrix:
symnum(Cl) # highly correlated

## Spearman's rho and Kendall's tau
symnum(clS <- cor(longley, method = "spearman"))
symnum(clK <- cor(longley, method = "kendall"))
## How much do they differ?
i <- lower.tri(Cl)
cor(cbind(P = Cl[i], S = clS[i], K = clK[i]))

## cov2cor() scales a covariance matrix by its diagonal
##           to become the correlation matrix.
cov2cor # see the function definition {and learn ..}
stopifnot(all.equal(Cl, cov2cor(cov(longley))),
          all.equal(cor(longley, method="kendall"),
            cov2cor(cov(longley, method="kendall"))))

## Missing value treatment:
data(swiss)
C1 <- cov(swiss)
range(eigen(C1, only=TRUE)$val) # 6.19  1921
swM <- swiss
swM[1,2] <- swM[7,3] <- swM[25,5] <- NA # create 3 "missing"
try(cov(swM)) # Error: missing obs...
C2 <- cov(swM, use = "complete")
range(eigen(C2, only=TRUE)$val) # 6.46  1930
```

```
C3 <- cov(swM, use = "pairwise")
range(eigen(C3, only=TRUE)$val) # 6.19  1938

(scM <- symnum(cor(swM, method = "kendall",
  use = "complete")))
## Kendall's tau doesn't change much: identical symnum
## codings!
identical(scM, symnum(cor(swiss, method = "kendall")))

all.equal(cov2cor(cov(swM, method = "kendall",
                      use = "pairwise")),
                  cor(swM, method = "kendall",
                      use = "pairwise"))
```

count.fields *Count the Number of Fields per Line*

Description

`count.fields` counts the number of fields, as separated by `sep`, in each of the lines of `file` read.

Usage

```
count.fields(file, sep = "", quote = "\"'", skip = 0,
             blank.lines.skip = TRUE, comment.char = "#")
```

Arguments

`file`	a character string naming an ASCII data file, or a `connection`, which will be opened if necessary, and if so closed at the end of the function call.
`sep`	the field separator character. Values on each line of the file are separated by this character. By default, arbitrary amounts of whitespace can separate fields.
`quote`	the set of quoting characters
`skip`	the number of lines of the data file to skip before beginning to read data.
`blank.lines.skip`	logical: if `TRUE` blank lines in the input are ignored.
`comment.char`	character: a character vector of length one containing a single character or an empty string.

Details

This used to be used by `read.table` and can still be useful in discovering problems in reading a file by that function.

For the handling of comments, see `scan`.

Value

A vector with the numbers of fields found.

See Also

`read.table`

Examples

```
cat("NAME", "1:John", "2:Paul", file = "foo", sep = "\n")
count.fields("foo", sep = ":")
unlink("foo")
```

cov.wt *Weighted Covariance Matrices*

Description

Returns a list containing estimates of the weighted covariance matrix and the mean of the data, and optionally of the (weighted) correlation matrix.

Usage

```
cov.wt(x, wt = rep(1/nrow(x), nrow(x)), cor = FALSE,
       center = TRUE)
```

Arguments

`x`	a matrix or data frame. As usual, rows are observations and columns are variables.
`wt`	a non-negative and non-zero vector of weights for each observation. Its length must equal the number of rows of `x`.
`cor`	A logical indicating whether the estimated correlation weighted matrix will be returned as well.
`center`	Either a logical or a numeric vector specifying the centers to be used when computing covariances. If `TRUE`, the (weighted) mean of each variable is used, if `FALSE`, zero is used. If `center` is numeric, its length must equal the number of columns of `x`.

Details

The covariance matrix is divided by one minus the sum of squares of the weights, so if the weights are the default $(1/n)$ the conventional unbiased estimate of the covariance matrix with divisor $(n - 1)$ is obtained. This differs from the behaviour in S-PLUS.

Value

A list containing the following named components:

`cov`	the estimated (weighted) covariance matrix
`center`	an estimate for the center (mean) of the data.

`n.obs` the number of observations (rows) in `x`.

`wt` the weights used in the estimation. Only returned if given as an argument.

`cor` the estimated correlation matrix. Only returned if `cor` is `TRUE`.

See Also

`cov` and `var`.

cut *Convert Numeric to Factor*

Description

cut divides the range of x into intervals and codes the values in x according to which interval they fall. The leftmost interval corresponds to level one, the next leftmost to level two and so on.

Usage

```
cut(x, ...)

## Default S3 method:
cut(x, breaks, labels = NULL,
    include.lowest = FALSE, right = TRUE, dig.lab = 3, ...)
```

Arguments

x	a numeric vector which is to be converted to a factor by cutting.
breaks	either a vector of cut points or number giving the number of intervals which x is to be cut into.
labels	labels for the levels of the resulting category. By default, labels are constructed using "(a,b]" interval notation. If labels = FALSE, simple integer codes are returned instead of a factor.
include.lowest	logical, indicating if an 'x[i]' equal to the lowest (or highest, for right = FALSE) 'breaks' value should be included.
right	logical, indicating if the intervals should be closed on the right (and open on the left) or vice versa.
dig.lab	integer which is used when labels are not given. It determines the number of digits used in formatting the break numbers.
...	further arguments passed to or from other methods.

Details

If a `labels` parameter is specified, its values are used to name the factor levels. If none is specified, the factor level labels are constructed as `"(b1, b2]"`, `"(b2, b3]"` etc. for `right=TRUE` and as `"[b1, b2)"`, ...if `right=FALSE`. In this case, `dig.lab` indicates how many digits should be used in formatting the numbers `b1`, `b2`, ...

Value

A `factor` is returned, unless `labels = FALSE` which results in the mere integer level codes.

Note

Instead of `table(cut(x, br))`, `hist(x, br, plot = FALSE)` is more efficient and less memory hungry. Instead of `cut(*, labels = FALSE)`, `findInterval()` is more efficient.

References

Becker, R. A., Chambers, J. M. and Wilks, A. R. (1988) *The New S Language.* Wadsworth & Brooks/Cole.

See Also

`split` for splitting a variable according to a group factor; `factor`, `tabulate`, `table`, `findInterval()`.

Examples

```
Z <- rnorm(10000)
table(cut(Z, br = -6:6))
sum(table(cut(Z, br = -6:6, labels=FALSE)))
sum(   hist  (Z, br = -6:6, plot=FALSE)$counts)

cut(rep(1,5),4) # dummy
tx0 <- c(9, 4, 6, 5, 3, 10, 5, 3, 5)
x <- rep(0:8, tx0)
stopifnot(table(x) == tx0)

table( cut(x, b = 8))
table( cut(x, br = 3*(-2:5)))
table( cut(x, br = 3*(-2:5), right = FALSE))
```

```
## some values OUTSIDE the breaks :
table(cx  <- cut(x, br = 2*(0:4)))
table(cxl <- cut(x, br = 2*(0:4), right = FALSE))
which(is.na(cx));  x[is.na(cx)]  # the first 9  values  0
which(is.na(cxl)); x[is.na(cxl)] # the last  5  values  8

## Label construction:
y <- rnorm(100)
table(cut(y, breaks = pi/3*(-3:3)))
table(cut(y, breaks = pi/3*(-3:3), dig.lab=4))

# extra digits don't "harm" here
table(cut(y, breaks =  1*(-3:3), dig.lab=4))
# the same, since no exact INT!
table(cut(y, breaks =  1*(-3:3), right = FALSE))
```

data.class *Object Classes*

Description

Determine the class of an arbitrary R object.

Usage

```
data.class(x)
```

Arguments

`x`	an R object.

Value

character string giving the "class" of `x`.

The "class" is the (first element) of the `class` attribute if this is non-`NULL`, or inferred from the object's `dim` attribute if this is non-`NULL`, or `mode(x)`.

Simply speaking, `data.class(x)` returns what is typically useful for method dispatching. (Or, what the basic creator functions already and maybe eventually all will attach as a class attribute.)

Note

For compatibility reasons, there is one exception to the rule above: When `x` is `integer`, the result of `data.class(x)` is `"numeric"` even when `x` is classed.

See Also

`class`

Examples

```
x <- LETTERS
data.class(factor(x))            # has a class attribute
data.class(matrix(x, nc = 13))   # has a dim attribute
data.class(list(x))              # the same as mode(x)
data.class(x)                    # the same as mode(x)
```

```
# compatibility "rule"
stopifnot(data.class(1:2) == "numeric")
```

data.frame *Data Frames*

Description

This function creates data frames, tightly coupled collections of variables which share many of the properties of matrices and of lists, used as the fundamental data structure by most of R's modeling software.

Usage

```
data.frame(..., row.names = NULL, check.rows = FALSE,
           check.names = TRUE)
```

Arguments

`...`	these arguments are of either the form `value` or `tag=value`. Component names are created based on the tag (if present) or the deparsed argument itself.
`row.names`	`NULL` or an integer or character string specifying a column to be used as row names, or a character vector giving the row names for the data frame.
`check.rows`	if `TRUE` then the rows are checked for consistency of length and names.
`check.names`	logical. If `TRUE` then the names of the variables in the data frame are checked to ensure that they are syntactically valid variable names. If necessary they are adjusted (by `make.names`) so that they are.

Details

A data frame is a list of variables of the same length with unique row names, given class `"data.frame"`.

`data.frame` converts each of its arguments to a data frame by calling `as.data.frame(optional=TRUE)`. As that is a generic function, methods can be written to change the behaviour of arguments according to their classes: R comes with many such methods. Character variables passed to `data.frame` are converted to factor columns unless protected by `I`. If a list or data frame or matrix is passed to `data.frame` it is as if each column had been passed as a separate argument.

Objects passed to `data.frame` should have the same number of rows, but atomic vectors, factors and character vectors protected by `I` will be recycled a whole number of times if necessary.

If row names are not supplied in the call to `data.frame`, the row names are taken from the first component that has suitable names, for example a named vector or a matrix with rownames or a data frame. (If that component is subsequently recycled, the names are discarded with a warning.) If `row.names` was supplied as `NULL` or no suitable component was found the row names are the integer sequence starting at one.

If row names are supplied of length one and the data frame has a single row, the `row.names` is taken to specify the row names and not a column (by name or number).

Value

A data frame, a matrix-like structure whose columns may be of differing types (numeric, logical, factor and character and so on).

Note

In versions of R prior to 1.4.0 logical columns were converted to factors (as in S3 but not S4).

References

Chambers, J. M. (1992) *Data for models.* Chapter 3 of *Statistical Models in S* eds J. M. Chambers and T. J. Hastie, Wadsworth & Brooks/Cole.

See Also

`I`, `plot.data.frame`, `print.data.frame`, `row.names`, `[.data.frame` for subsetting methods, `Math.data.frame` etc, about *Group* methods for `data.frames`; `read.table`, `make.names`.

Examples

```
L3 <- LETTERS[1:3]
str(d <- data.frame(cbind(x=1, y=1:10),
                    fac=sample(L3, 10, repl=TRUE)))

## The same with automatic column names:
str(data.frame(cbind( 1, 1:10), sample(L3, 10, repl=TRUE)))
is.data.frame(d)
```

```
## do not convert to factor, using I() :
str(cbind(d, char = I(letters[1:10])), vec.len = 10)

stopifnot(1:10 == row.names(d)) # {coercion}

(d0  <- d[, FALSE]) # NULL data frame with 10 rows
(d.0 <- d[FALSE, ]) # <0 rows> data frame (3 cols)
(d00 <- d0[FALSE,]) # NULL data frame with 0 rows
```

`data.matrix` *Data Frame to Numeric Matrix*

Description

Return the matrix obtained by converting all the variables in a data frame to numeric mode and then binding them together as the columns of a matrix. Factors and ordered factors are replaced by their internal codes.

Usage

```
data.matrix(frame)
```

Arguments

`frame`	a data frame whose components are logical vectors, factors or numeric vectors.

References

Chambers, J. M. (1992) *Data for models.* Chapter 3 of *Statistical Models in S* eds J. M. Chambers and T. J. Hastie, Wadsworth & Brooks/Cole.

See Also

`as.matrix`, `data.frame`, `matrix`.

`dataentry` *Spreadsheet Interface for Entering Data*

Description

A spreadsheet-like editor for entering or editing data.

Usage

```
data.entry(..., Modes = NULL, Names = NULL)
dataentry(data, modes)
de(..., Modes = list(), Names = NULL)
```

Arguments

`...`	A list of variables: currently these should be numeric or character vectors or list containing such vectors.
`Modes`	The modes to be used for the variables.
`Names`	The names to be used for the variables.
`data`	A list of numeric and/or character vectors.
`modes`	A list of length up to that of `data` giving the modes of (some of) the variables. `list()` is allowed.

Details

The data entry editor is only available on some platforms and GUIs. Where available it provides a means to visually edit a matrix or a collection of variables (including a data frame) as described in the "Notes" section.

`data.entry` has side effects, any changes made in the spreadsheet are reflected in the variables. The functions `de`, `de.ncols`, `de.setup` and `de.restore` are designed to help achieve these side effects. If the user passes in a matrix, `X` say, then the matrix is broken into columns before `dataentry` is called. Then on return the columns are collected and glued back together and the result assigned to the variable `X`. If you don't want this behaviour, use data.entry directly.

The primitive function is `dataentry`. It takes a list of vectors of possibly different lengths and modes (the second argument) and opens a spreadsheet with these variables being the columns. The columns of the dataentry window are returned as vectors in a list when the spreadsheet is closed.

`de.ncols` counts the number of columns which are supplied as arguments to `data.entry`. It attempts to count columns in lists, matrices and vectors. `de.setup` sets things up so that on return the columns can be regrouped and reassigned to the correct name. This is handled by `de.restore`.

Value

`de` and `dataentry` return the edited value of their arguments. `data.entry` invisibly returns a vector of variable names but its main value is its side effect of assigning new version of those variables in the user's workspace.

Note

The details of interface to the data grid may differ by platform and GUI. The following description applies to the X11-based implementation under Unix.

You can navigate around the grid using the cursor keys or by clicking with the (left) mouse button on any cell. The active cell is highlighted by thickening the surrounding rectangle. Moving to the right or down will scroll the grid as needed: there is no constraint to the rows or columns currently in use.

There are alternative ways to navigate using the keys. Return and (keypad) Enter and LineFeed all move down. Tab moves right and Shift-Tab move left. Home moves to the top left.

PageDown or Control-F moves down a page, and PageUp or Control-B up by a page. End will show the last used column and the last few rows used (in any column).

Using any other key starts an editing process on the currently selected cell: moving away from that cell enters the edited value whereas Esc cancels the edit and restores the previous value. When the editing process starts, the cell is cleared. In numerical columns (the default) only letters making up a valid number (including `-.eE`) are accepted, and entering an invalid edited value (such as blank) enters `NA` in that cell. The last entered value can be deleted using the BackSpace or Del(ete) key. Only a limited number of characters (currently 29) can be entered in a cell, and if necessary only the start or end of the string will be displayed, with the omissions indicated by > or <. (The start is shown except when editing.)

Entering a value in a cell further down a column than the last used cell extends the variable and fills the gap (if any) by `NA`s (not shown on screen).

The column names can only be selected by clicking in them. This gives a popup menu to select the column type (currently Real (numeric) or Character) or to change the name. Changing the type converts the current contents of the column (and converting from Character to Real may generate `NA`s.) If changing the name is selected the header cell becomes editable (and is cleared). As with all cells, the value is entered by moving away from the cell by clicking elsewhere or by any of the keys for moving down (only).

New columns are created by entering values in them (and not by just assigning a new name). The mode of the column is auto-detected from the first value entered: if this is a valid number it gives a numeric column. Unused columns are ignored, so adding data in `var5` to a three-column grid adds one extra variable, not two.

The `Copy` button copies the currently selected cell: `paste` copies the last copied value to the current cell, and right-clicking selects a cell *and* copies in the value. Initially the value is blank, and attempts to paste a blank value will have no effect.

Control-L will refresh the display, recalculating field widths to fit the current entries.

In the default mode the column widths are chosen to fit the contents of each column, with a default of 10 characters for empty columns. you can specify fixed column widths by setting option `de.cellwidth` to the required fixed width (in characters). (set it to zero to return to variable widths). The displayed width of any field is limited to 600 pixels (and by the window width).

See Also

`vi`, `edit`: `edit` uses `dataentry` to edit data frames.

Examples

```
# call data entry with variables x and y
data.entry(x,y)
```

dataframeHelpers *Data Frame Auxiliary Functions*

Description

Auxiliary functions for use with data frames.

Usage

```
xpdrows.data.frame(x, old.rows, new.rows)
```

Arguments

`x`	object of class `data.frame`.
`old.rows, new.rows`	row names for old and new rows.

Details

`xpdrows.data.frame` is an auxiliary function which expands the rows of a data frame. It is used by the data frame methods of `[<-` and `[[<-` (which perform subscripted assignments on a data frame), and not intended to be called directly.

See Also

`[.data.frame`

date *System Date and Time*

Description

Returns a character string of the current system date and time.

Usage

```
date()
```

Value

The string has the form `"Fri Aug 20 11:11:00 1999"`, i.e., length 24, since it relies on POSIX' `ctime` ensuring the above fixed format. Timezone and Daylight Saving Time are taken account of, but *not* indicated in the result.

References

Becker, R. A., Chambers, J. M. and Wilks, A. R. (1988) *The New S Language*. Wadsworth & Brooks/Cole.

Examples

```
(d <- date())
nchar(d) == 24
```

dcf *Read and Write Data in DCF Format*

Description

Reads or writes an R object from/to a file in Debian Control File format.

Usage

```
read.dcf(file, fields=NULL)
write.dcf(x, file = "", append = FALSE,
          indent = 0.1 * getOption("width"),
          width = 0.9 * getOption("width"))
```

Arguments

`file`	either a character string naming a file or a connection. `""` indicates output to the console.
`fields`	Fields to read from the DCF file. Default is to read all fields.
`x`	the object to be written, typically a data frame. If not, it is attempted to coerce `x` to a data frame.
`append`	logical. If `TRUE`, the output is appended to the file. If `FALSE`, any existing file of the name is destroyed.
`indent`	a positive integer specifying the indentation for continuation lines in output entries.
`width`	a positive integer giving the target column for wrapping lines in the output.

Details

DCF is a simple format for storing databases in plain text files that can easily be directly read and written by humans. DCF is used in various places to store R system information, like descriptions and contents of packages.

The DCF rules as implemented in R are:

1. A database consists of one or more records, each with one or more named fields. Not every record must contain each field, a field may appear only once in a record.
2. Regular lines start with a non-whitespace character.

3. Regular lines are of form `tag:value`, i.e., have a name tag and a value for the field, separated by : (only the first : counts). The value can be empty (=whitespace only).
4. Lines starting with whitespace are continuation lines (to the preceding field) if at least one character in the line is non-whitespace.
5. Records are separated by one or more empty (=whitespace only) lines.

`read.dcf` returns a character matrix with one line per record and one column per field. Leading and trailing whitespace of field values is ignored. If a tag name is specified, but the corresponding value is empty, then an empty string of length 0 is returned. If the tag name of a fields is never used in a record, then `NA` is returned.

See Also

`write.table`.

Examples

```
## Create a reduced version of the 'CONTENTS' file in
## package 'eda'
x <- read.dcf(file = system.file("CONTENTS",
                                 package = "eda"),
              fields = c("Entry", "Description"))
write.dcf(x)
```

debug *Debug a function*

Description

Set or unset the debugging flag on a function.

Usage

```
debug(fun)
undebug(fun)
```

Arguments

`fun`	any interpreted R function.

Details

When a function flagged for debugging is entered, normal execution is suspended and the body of function is executed one statement at a time. A new browser context is initiated for each step (and the previous one destroyed). Currently you can only debug functions that have bodies enclosed in braces. This is a bug and will be fixed soon. You take the next step by typing carriage return, `n` or `next`. You can see the values of variables by typing their names. Typing `c` or `cont` causes the debugger to continue to the end of the function. You can `debug` new functions before you step in to them from inside the debugger. Typing `Q` quits the current execution and returns you to the top–level prompt. Typing `where` causes the debugger to print out the current stack trace (all functions that are active). If you have variables with names that are identical to the controls (eg. `c` or `n`) then you need to use `print(c)` and `print(n)` to evaluate them.

See Also

`browser`, `traceback` to see the stack after an `Error: ...` message; `recover` for another debugging approach.

`debugger` *Post-Mortem Debugging*

Description

Functions to dump the evaluation environments (frames) and to examine dumped frames.

Usage

```
dump.frames(dumpto = "last.dump", to.file = FALSE)
debugger(dump = last.dump)
```

Arguments

`dumpto`	a character string. The name of the object or file to dump to.
`to.file`	logical. Should the dump be to an R object or to a file?
`dump`	An R dump object created by `dump.frames`.

Details

To use post-mortem debugging, set the option `error` to be a call to `dump.frames`. By default this dumps to an R object `"last.dump"` in the workspace, but it can be set to dump to a file (as dump of the object produced by a call to `save`). The dumped object contain the call stack, the active environments and the last error message as returned by `geterrmessage`.

When dumping to file, `dumpto` gives the name of the dumped object and the file name has `.rda` appended.

A dump object of class `"dump.frames"` can be examined by calling `debugger`. This will give the error message and a list of environments from which to select repeatedly. When an environment is selected, it is copied and the `browser` called from within the copy.

If `dump.frames` is installed as the error handler, execution will continue even in non-interactive sessions. See the examples for how to dump and then quit.

Value

None.

Note

Functions such as **sys.parent** and **environment** applied to closures will not work correctly inside **debugger**.

Of course post-mortem debugging will not work if R is too damaged to produce and save the dump, for example if it has run out of workspace.

References

Becker, R. A., Chambers, J. M. and Wilks, A. R. (1988) *The New S Language.* Wadsworth & Brooks/Cole.

See Also

options for setting **error** options; **recover** is an interactive debugger working similarly to **debugger** but directly after the error occurs.

Examples

```
options(error=quote(dump.frames("testdump", TRUE)))

f <- function() {
    g <- function() stop("test dump.frames")
    g()
}
f()   # will generate a dump on file "testdump.rda"
options(error=NULL)

## possibly in another R session
load("testdump.rda")
debugger(testdump)
Available environments had calls:
1: f()
2: g()
3: stop("test dump.frames")

Enter an environment number, or 0 to exit
Selection: 1
Browsing in the environment with call:
f()
Called from: debugger.look(ind)
Browse[1]> ls()
[1] "g"
Browse[1]> g
```

```
function() stop("test dump.frames")
<environment: 759818>
Browse[1]>
Available environments had calls:
1: f()
2: g()
3: stop("test dump.frames")

Enter an environment number, or 0 to exit
Selection: 0

## A possible setting for non-interactive sessions
options(error=quote({dump.frames(to.file=TRUE); q()}))
```

Defunct *Defunct Functions*

Description

The functions or variables listed here are no longer part of R as they are not needed (any more).

Usage

```
.Defunct()

Version()
provide(package)
.Provided
category(...)
dnchisq(.)
pnchisq(.)
qnchisq(.)
rnchisq(.)
print.anova.glm(.)
print.anova.lm(.)
print.tabular(.)
print.plot(.)
save.plot(.)
system.test(.)
dotplot(...)
stripplot(...)
getenv(...)
read.table.url(url, method,...)
scan.url(url, file = tempfile(), method, ...)
source.url(url, file = tempfile(), method, ...)
httpclient(url, port=80, error.is.fatal=TRUE,
           check.MIME.type=TRUE, file=tempfile(),
           drop.ctrl.z=TRUE)
parse.dcf(text = NULL, file = "", fields = NULL,
          versionfix = FALSE)
.Alias(expr)
reshapeWide(x, i = reshape.i, j = reshape.j,
            val = reshape.v, jnames = levels(j))
reshapeLong(x,jvars,  ilev = row.names(x),
            jlev = names(x)[jvars], iname = "reshape.i",
```

```
             jname = "reshape.j", vname = "reshape.v")
piechart(x, labels = names(x), edges = 200, radius = 0.8,
         density = NULL, angle = 45, col = NULL,
         main = NULL, ...)
print.ordered(.)
.Dyn.libs
.lib.loc
machine()
Machine()
Platform()
restart()
printNoClass(x, digits = NULL, quote = TRUE,
             na.print = NULL, print.gap = NULL,
             right = FALSE, ...)
plot.mts(x, plot.type = c("multiple", "single"),
         panel = lines, log = "", col = par("col"),
         bg = NA, pch = par("pch"), cex = par("cex"),
         lty = par("lty"), lwd = par("lwd"),
         ann = par("ann"),  xlab = "Time", type = "l",
         main=NULL, oma=c(6, 0, 5, 0), ...)
```

Details

`.Defunct` is the function to which defunct functions are set.

`category` has been an old-S function before there were factors; should be replaced by `factor` throughout!

The `*chisq()` functions now take an optional non-centrality argument, so the `*nchisq()` functions are no longer needed.

The new function `dev.print()` should now be used for saving plots to a file or printing them.

`provide` and its object `.Provided` have been removed. They were never used for their intended purpose, to allow one package to subsume another.

`dotplot` and `stripplot` have been renamed to `dotchart` and `stripchart`, respectively.

`getenv` has been replaced by `Sys.getenv`.

`*.url` are replaced by calling `read.table`, `scan` or `source` on a `url` connection.

`httpclient` was used by the deprecated `"socket"` method of `download.file`.

`parse.dcf` has been replaced by `read.dcf`, which is much faster, but has a slightly different interface.

`.Alias` provided an unreliable way to create duplicate references to the same object. There is no direct replacement. Where multiple references to a single object are required for semantic reasons consider using environments or external pointers. There are some notes on `http://developer.r-project.org`.

`reshape*`, which were experimental, are replaced by `reshape`. This has a different syntax and allows multiple time-varying variables.

`piechart` is the old name for `pie`, but clashed with usage in Trellis.

`.Dyn.libs` and `.lib.loc` were internal variables used for storing and manipulating the information about packages with dynloaded shared libs, and the known R library trees. These are now dynamic variables which one can get or set using `.dynLibs` and `.libPaths`, respectively.

`Machine()` and `Platform()` were functions returning the variables `.Machine` and `.Platform` respectively.

`restart()` should be replaced by `try()`, in preparation for an exception-based implementation. If you use `restart()` in a way that cannot be replaced with `try()` then ask for help on `r-devel`.

`printNoClass` was in package **methods** and calls directly the internal function `print.default`.

`plot.mts` has been removed, as `plot.ts` now has the same functionality.

See Also

`Deprecated`

delay *Delay Evaluation*

Description

delay creates a *promise* to evaluate the given expression in the specified environment if its value is requested. This provides direct access to *lazy evaluation* mechanism used by R for the evaluation of (interpreted) functions.

Usage

```
delay(x, env=.GlobalEnv)
```

Arguments

x	an expression.
env	an evaluation environment

Details

This is an experimental feature and its addition is purely for evaluation purposes.

Value

A *promise* to evaluate the expression. The value which is returned by delay can be assigned without forcing its evaluation, but any further accesses will cause evaluation.

Examples

```
x <- delay({
    for(i in 1:7)
        cat("yippee!\n")
    10
})

x^2 # yippee
x^2 # simple number
```

delete.response *Modify Terms Objects*

Description

delete.response returns a terms object for the same model but with no response variable.

drop.terms removes variables from the right-hand side of the model. There is also a "[.terms" method to perform the same function (with keep.response=TRUE).

reformulate creates a formula from a character vector.

Usage

```
delete.response(termobj)
reformulate(termlabels, response = NULL)
drop.terms(termobj, dropx = NULL, keep.response = FALSE)
```

Arguments

termobj	A terms object
termlabels	character vector giving the right-hand side of a model formula.
response	character string, symbol or call giving the left-hand side of a model formula.
dropx	vector of positions of variables to drop from the right-hand side of the model.
keep.response	Keep the response in the resulting object?

Value

delete.response and drop.terms return a terms object.

reformulate returns a formula.

See Also

terms

Examples

```
ff <- y ~ z + x + w
tt <- terms(ff)
tt
delete.response(tt)
drop.terms(tt, 2:3, keep.response = TRUE)
tt[-1]
tt[2:3]
reformulate(attr(tt, "term.labels"))

## keep LHS :
reformulate("x*w", ff[[2]])
fS <- surv(ft, case) ~ a + b
reformulate(c("a", "b*f"), fS[[2]])

stopifnot(identical( ~ var, reformulate("var")),
  identical(~ a + b + c, reformulate(letters[1:3])),
  identical(  y ~ a + b, reformulate(letters[1:2], "y"))
)
```

demo *Demonstrations of R Functionality*

Description

demo is a user-friendly interface to running some demonstration R scripts. demo() gives the list of available topics.

Usage

```
demo(topic, device = getOption("device"),
     package = .packages(), lib.loc = NULL,
     character.only = FALSE, verbose = getOption("verbose"))
```

Arguments

topic	the topic which should be demonstrated, given as a name or literal character string, or a character string, depending on whether character.only is FALSE (default) or TRUE. If omitted, the list of available topics is displayed.
device	the graphics device to be used.
package	a character vector giving the packages to look into for demos. By default, all packages in the search path are used.
lib.loc	a character vector of directory names of R libraries, or NULL. The default value of NULL corresponds to all libraries currently known. If the default is used, the loaded packages are searched before the libraries.
character.only	logical; if TRUE, use topic as character string.
verbose	a logical. If TRUE, additional diagnostics are printed.

Details

If no topics are given, demo lists the available demos. The corresponding information is returned in an object of class "packageIQR". The structure of this class is experimental. In earlier versions of R, an empty character vector was returned along with listing available demos.

See Also

`source` which is called by `demo`.

Examples

```
demo() # for attached packages

## All available demos:
demo(package = .packages(all.available = TRUE))

demo(lm.glm)

ch <- "scoping"
demo(ch, character = TRUE)
```

deparse *Expression Deparsing*

Description

Turn unevaluated expressions into character strings.

Usage

```
deparse(expr, width.cutoff = 60,
  backtick = mode(expr) %in% c("call", "expression", "("))
```

Arguments

`expr`	any R expression.
`width.cutoff`	integer in $[20, 500]$ determining the cutoff at which line-breaking is tried.
`backtick`	logical indicating whether symbolic names should be enclosed in backticks if they don't follow the standard syntax.

Details

This function turns unevaluated expressions (where "expression" is taken in a wider sense than the strict concept of a vector of mode `"expression"` used in `expression`) into character strings (a kind of inverse `parse`).

A typical use of this is to create informative labels for data sets and plots. The example shows a simple use of this facility. It uses the functions `deparse` and `substitute` to create labels for a plot which are character string versions of the actual arguments to the function `myplot`.

The default for the `backtick` option is not to quote single symbols but only composite expressions. This is a compromise to avoid breaking existing code.

References

Becker, R. A., Chambers, J. M. and Wilks, A. R. (1988) *The New S Language*. Wadsworth & Brooks/Cole.

See Also

`substitute`, `parse`, `expression`.

Examples

```
deparse(args(lm))
deparse(args(lm), width = 500)
myplot <-
function(x, y)
    plot(x, y, xlab=deparse(substitute(x)),
        ylab=deparse(substitute(y)))
```

Deprecated *Deprecated Functions*

Description

These functions are provided for compatibility with older versions of R only, and may be defunct as soon as of the next release.

Usage

```
.Deprecated(new)

print.coefmat(x, digits=max(3, getOption("digits") - 2),
  signif.stars = getOption("show.signif.stars"),
  dig.tst = max(1, min(5, digits - 1)),
  cs.ind = 1:k, tst.ind = k + 1, zap.ind = integer(0),
  P.values = NULL,
  has.Pvalue = nc >= 4 &&
    substr(colnames(x)[nc],1,3) == "Pr(",
  eps.Pvalue = .Machine$double.eps,
  na.print = "", ...)

codes(x, ...)
codes(x, ...) <- value

anovalist.lm(object, ..., test = NULL)
lm.fit.null(x, y, method = "qr", tol = 1e-07, ...)
lm.wfit.null(x, y, w, method = "qr", tol = 1e-07, ...)
glm.fit.null(x, y, weights = rep(1, nobs), start = NULL,
  etastart = NULL, mustart = NULL, offset = rep(0, nobs),
  family = gaussian(), control = glm.control(),
  intercept = FALSE)
print.atomic(x, quote = TRUE, ...)
```

Details

`.Deprecated("<new name>")` is called from deprecated functions. The original help page for these functions is often available at `help("oldName-deprecated")` (note the quotes).

`tkfilefind` is a demo in package **tcltk** displaying a widget for selecting files but the same functionality is available in a better form in the `tkgetOpenFile` and `tkgetSaveFile` functions. The demo is reported not even to work with recent versions of Tcl and Tk libraries.

`print.coefmat` is an older name for `printCoefmat` with a different default for `na.print`.

`codes` was almost always used inappropriately. To get the internal coding of a factor, use `unclass`, `as.vector` or `as.integer`. For *ordered* factors, `codes` was equivalent to these, but for *unordered* factors it assumed an an alphabetical ordering of the levels in the locale in use.

`anovalist.lm` was replaced by `anova.lmlist` in R 1.2.0.

`lm.fit.null` and `lm.wfit.null` are superseded by `lm.fit` and `lm.wfit` which handle null models now. Similarly, `glm.fit.null` is superseded by `glm.fit`.

`print.atomic` differs from `print.default` only in its argument sequence. It is not a method for `print`.

See Also

`Defunct`,

det *Calculate the Determinant of a Matrix*

Description

`det` calculates the determinant of a matrix. `determinant` is a generic function that returns separately the modulus of the determinant, optionally on the logarithm scale, and the sign of the determinant.

Usage

```
det(x, ...)
determinant(x, logarithm = TRUE, ...)
```

Arguments

`x`	numeric matrix.
`logarithm`	logical; if `TRUE` (default) return the logarithm of the modulus of the determinant.
`...`	Optional arguments. At present none are used. Previous versions of `det` allowed an optional `method` argument. This argument will be ignored but will not produce an error.

Value

For `det`, the determinant of `x`. For `determinant`, a list with components

`modulus`	a numeric value. The modulus (absolute value) of the determinant if `logarithm` is `FALSE`; otherwise the logarithm of the modulus.
`sign`	integer; either $+1$ or -1 according to whether the determinant is positive or negative.

Note

Often, computing the determinant is *not* what you should be doing to solve a given problem.

Prior to version 1.8.0 the `det` function had a `method` argument to allow use of either a QR decomposition or an eigenvalue-eigenvector decomposition. The `determinant` function now uses an LU decomposition and the `det` function is simply a wrapper around a call to `determinant`.

Examples

```
(x <- matrix(1:4, ncol=2))
unlist(determinant(x))
det(x)

det(print(cbind(1,1:3,c(2,0,1))))
```

detach *Detach Objects from the Search Path*

Description

Detach a database, i.e., remove it from the `search()` path of available R objects. Usually, this is either a `data.frame` which has been `attached` or a package which was required previously.

Usage

```
detach(name, pos = 2, version)
```

Arguments

`name`	The object to detach. Defaults to `search()[pos]`. This can be a name or a character string but *not* a character vector.
`pos`	Index position in `search()` of database to detach. When `name` is `numeric`, `pos = name` is used.
`version`	A character string denoting a version number of the package to be loaded. If no version is given, a suitable default is chosen.

Value

The attached database is returned invisibly, either as `data.frame` or as `list`.

Note

You cannot detach either the workspace (position 1) or the **base** package (the last item in the search list).

References

Becker, R. A., Chambers, J. M. and Wilks, A. R. (1988) *The New S Language*. Wadsworth & Brooks/Cole.

See Also

`attach`, `library`, `search`, `objects`.

Examples

```
require(eda) # package
detach(package:eda)
## could equally well use detach("package:eda") but NOT
## pkg <- "package:eda"; detach(pkg) Instead, use
library(eda)
pkg <- "package:eda"
detach(pos = match(pkg, search()))

## careful: do not do this unless 'lqs' is not already
## loaded.
library(lqs)
detach(2) # 'pos' used for 'name'
```

diag *Matrix Diagonals*

Description

Extract or replace the diagonal of a matrix, or construct a diagonal matrix.

Usage

```
diag(x = 1, nrow, ncol= )
diag(x) <- value
```

Arguments

`x`	a matrix, vector or 1D array.
`nrow, ncol`	Optional dimensions for the result.
`value`	either a single value or a vector of length equal to that of the current diagonal. Should be of a mode which can be coerced to that of `x`.

Value

If `x` is a matrix then `diag(x)` returns the diagonal of `x`. The resulting vector will have `names` if the matrix `x` has matching column and row names.

If `x` is a vector (or 1D array) of length two or more, then `diag(x)` returns a diagonal matrix whose diagonal is `x`.

If `x` is a vector of length one then `diag(x)` returns an identity matrix of order the nearest integer to `x`. The dimension of the returned matrix can be specified by `nrow` and `ncol` (the default is square).

The assignment form sets the diagonal of the matrix `x` to the given value(s).

Note

Using `diag(x)` can have unexpected effects if `x` is a vector that could be of length one. Use `diag(x, nrow = length(x))` for consistent behaviour.

References

Becker, R. A., Chambers, J. M. and Wilks, A. R. (1988) *The New S Language.* Wadsworth & Brooks/Cole.

See Also

`upper.tri`, `lower.tri`, `matrix`.

Examples

```
dim(diag(3))
diag(10,3,4) # guess what?
all(diag(1:3) == {m <- matrix(0,3,3); diag(m) <- 1:3; m})

# vector with names "X" and "Y"
diag(var(M <- cbind(X=1:5, Y=rnorm(5))))
rownames(M) <- c(colnames(M),rep("",3));
M; diag(M) # named as well
```

dim *Dimensions of an Object*

Description

Retrieve or set the dimension of an object.

Usage

```
dim(x)
dim(x) <- value
```

Arguments

x	an R object, for example a matrix, array or data frame.
value	For the default method, either NULL or a numeric vector which is coerced to integer (by truncation).

Details

The functions dim and dim<- are generic.

dim has a method for data.frames, which returns the length of the row.names attribute of x and the length of x (the numbers of "rows" and "columns").

Value

For an array (and hence in particular, for a matrix) dim retrieves the dim attribute of the object. It is NULL or a vector of mode integer.

References

Becker, R. A., Chambers, J. M. and Wilks, A. R. (1988) *The New S Language*. Wadsworth & Brooks/Cole.

See Also

ncol, nrow and dimnames.

Examples

```
x <- 1:12 ; dim(x) <- c(3,4)
x

# simple versions of nrow and ncol could be defined as
# follows
nrow0 <- function(x) dim(x)[1]
ncol0 <- function(x) dim(x)[2]
```

dimnames *Dimnames of an Object*

Description

Retrieve or set the dimnames of an object.

Usage

```
dimnames(x)
dimnames(x) <- value
```

Arguments

`x`	an R object, for example a matrix, array or data frame.
`value`	a possible value for `dimnames(x)`: see "Value".

Details

The functions `dimnames` and `dimnames<-` are generic.

For an `array` (and hence in particular, for a `matrix`), they retrieve or set the `dimnames` attribute (see attributes) of the object. The list `value` can have names, and these will be used to label the dimensions of the array where appropriate.

Both have methods for data frames. The dimnames of a data frame are its `row.names` attribute and its `names`.

As from R 1.8.0 factor components of `value` will be coerced to character.

Value

The dimnames of a matrix or array can be `NULL` or a list of the same length as `dim(x)`. If a list, its components are either `NULL` or a character vector the length of the appropriate dimension of `x`.

References

Becker, R. A., Chambers, J. M. and Wilks, A. R. (1988) *The New S Language.* Wadsworth & Brooks/Cole.

See Also

`rownames`, `colnames`; `array`, `matrix`, `data.frame`.

Examples

```
## simple versions of rownames and colnames could be
## defined as follows
rownames0 <- function(x) dimnames(x)[[1]]
colnames0 <- function(x) dimnames(x)[[2]]
```

`do.call` *Execute a Function Call*

Description

`do.call` executes a function call from the name of the function and a list of arguments to be passed to it.

Usage

```
do.call(what, args)
```

Arguments

`what`	a character string naming the function to be called.
`args`	a *list* of arguments to the function call. The `names` attribute of `args` gives the argument names.

Value

The result of the (evaluated) function call.

References

Becker, R. A., Chambers, J. M. and Wilks, A. R. (1988) *The New S Language*. Wadsworth & Brooks/Cole.

See Also

`call` which creates an unevaluated call.

Examples

```
do.call("complex", list(imag = 1:3))
```

double *Double Precision Vectors*

Description

Create, coerce to or test for a double-precision vector.

Usage

```
double(length = 0)
as.double(x, ...)
is.double(x)
single(length = 0)
as.single(x, ...)
```

Arguments

`length`	desired length.
`x`	object to be coerced or tested.
`...`	further arguments passed to or from other methods.

Value

`double` creates a double precision vector of the specified length. The elements of the vector are all equal to `0`.

`as.double` attempts to coerce its argument to be of double type: like `as.vector` it strips attributes including names.

`is.double` returns `TRUE` or `FALSE` depending on whether its argument is of double type or not. It is generic: you can write methods to handle specific classes of objects, see InternalMethods.

Note

R *has no single precision data type. All real numbers are stored in double precision format.* The functions `as.single` and `single` are identical to `as.double` and `double` except they set the attribute `Csingle` that is used in the `.C` and `.Fortran` interface, and they are intended only to be used in that context.

References

Becker, R. A., Chambers, J. M. and Wilks, A. R. (1988) *The New S Language.* Wadsworth & Brooks/Cole.

See Also

`integer`.

Examples

```
is.double(1)
all(double(3) == 0)
```

`download.file` *Download File from the Internet*

Description

This function can be used to download a file from the Internet.

Usage

```
download.file(url, destfile, method, quiet = FALSE,
              mode="w", cacheOK = TRUE)
```

Arguments

`url`	A character string naming the URL of a resource to be downloaded.
`destfile`	A character string with the name where the downloaded file is saved. Tilde-expansion is performed.
`method`	Method to be used for downloading files. Currently download methods `"internal"`, `"wget"` and `"lynx"` are available. The default is to choose the first of these which will be `"internal"`. The method can also be set through the option `"download.file.method"`: see `options()`.
`quiet`	If `TRUE`, suppress status messages (if any).
`mode`	character. The mode with which to write the file. Useful values are `"w"`, `"wb"` (binary), `"a"` (append) and `"ab"`. Only used for the `"internal"` method.
`cacheOK`	logical. Is a server-side cached value acceptable? Implemented for the `"internal"` and `"wget"` methods.

Details

The function `download.file` can be used to download a single file as described by `url` from the internet and store it in `destfile`. The `url` must start with a scheme such as `"http://"`, `"ftp://"` or `"file://"`.

`cacheOK = FALSE` is useful for `"http://"` URLs, and will attempt to get a copy directly from the site rather than from an intermediate cache. (Not all platforms support it.) It is used by `CRAN.packages`.

The remaining details apply to method `"internal"` only.

The timeout for many parts of the transfer can be set by the option `timeout` which defaults to 60 seconds.

The level of detail provided during transfer can be set by the `quiet` argument and the `internet.info` option. The details depend on the platform and scheme, but setting `internet.info` to 0 gives all available details, including all server responses. Using 2 (the default) gives only serious messages, and 3 or more suppresses all messages.

Method `"wget"` can be used with proxy firewalls which require user/password authentication if proper values are stored in the configuration file for `wget`.

Setting Proxies

This applies to the internal code only.

Proxies can be specified via environment variables. Setting `"no_proxy"` stops any proxy being tried. Otherwise the setting of `"http_proxy"` or `"ftp_proxy"` (or failing that, the all upper-case version) is consulted and if non-empty used as a proxy site. For FTP transfers, the username and password on the proxy can be specified by `"ftp_proxy_user"` and `"ftp_proxy_password"`. The form of `"http_proxy"` should be `"http://proxyhost/"` or `"http://proxyhost:8080/"` where the port defaults to 80 and the trailing slash may be omitted. For `"ftp_proxy"` use the form `"ftp://proxyhost:3128/"` where the default port is 21. These environment variables must be set before the download code is first used: they cannot be altered later by calling `Sys.putenv`.

Usernames and passwords can be set for HTTP proxy transfers via environment variable `http_proxy_user` in the form `user:passwd`. Alternatively, `"http_proxy"` can be of the form `"http://user:pass@proxy.dom.com:8080/"` for compatibility with `wget`. Only the HTTP/1.0 basic authentication scheme is supported.

Note

Methods `"wget"` and `"lynx"` are for historical compatibility. They will block all other activity on the R process.

For methods `"wget"` and `"lynx"` a system call is made to the tool given by `method`, and the respective program must be installed on your system and be in the search path for executables.

See Also

`options` to set the `timeout` and `internet.info` options.

`url` for a finer-grained way to read data from URLs.

`url.show`, `CRAN.packages`, `download.packages` for applications

dput *Write an Internal Object to a File*

Description

Writes an ASCII text representation of an R object to a file or connection, or uses one to recreate the object.

Usage

```
dput(x, file = "")
dget(file)
```

Arguments

`x`	an object.
`file`	either a character string naming a file or a connection. `""` indicates output to the console.

Details

`dput` opens `file` and deparses the object `x` into that file. The object name is not written (contrary to `dump`). If `x` is a function the associated environment is stripped. Hence scoping information can be lost.

Using `dget`, the object can be recreated (with the limitations mentioned above).

`dput` will warn if fewer characters were written to a file than expected, which may indicate a full or corrupt file system.

References

Becker, R. A., Chambers, J. M. and Wilks, A. R. (1988) *The New S Language*. Wadsworth & Brooks/Cole.

See Also

`deparse`, `dump`, `write`.

Examples

```
## Write an ASCII version of mean to the file "foo"
dput(mean, "foo")
## And read it back into 'bar'
bar <- dget("foo")
unlink("foo")
```

drop *Drop Redundant Extent Information*

Description

Delete the dimensions of an array which have only one level.

Usage

```
drop(x)
```

Arguments

x	an array (including a matrix).

Value

If x is an object with a dim attribute (e.g., a matrix or array), then drop returns an object like x, but with any extents of length one removed. Any accompanying dimnames attribute is adjusted and returned with x.

Array subsetting ([) performs this reduction unless used with drop = FALSE, but sometimes it is useful to invoke drop directly.

See Also

drop1 which is used for dropping terms in models.

Examples

```
dim(drop(array(1:12, dim=c(1,3,1,1,2,1,2)))) # = 3 2 2
drop(1:3 %*% 2:4) # scalar product
```

dump *Text Representations of R Objects*

Description

This function takes a vector of names of R objects and produces text representations of the objects on a file or connection. A dump file can be `source`d into another R (or S) session.

Usage

```
dump(list, file = "dumpdata.R", append = FALSE,
     envir = parent.frame())
```

Arguments

`list`	character. The names of one or more R objects to be dumped.
`file`	either a character string naming a file or a connection. `""` indicates output to the console.
`append`	if `TRUE`, output will be appended to `file`; otherwise, it will overwrite the contents of `file`.
`envir`	the environment to search for objects.

Details

At present the implementation of dump is very incomplete and it really only works for functions and simple vectors.

dump will warn if fewer characters were written to a file than expected, which may indicate a full or corrupt file system.

The function `save` is designed to be used for transporting R data between machines.

Note

The `envir` argument was added at version 1.7.0, and changed the search path for named objects to include the environment from which dump was called.

As dump is defined in the base namespace, the **base** package will be searched *before* the global environment unless dump is called from the top level or the `envir` argument is given explicitly.

References

Becker, R. A., Chambers, J. M. and Wilks, A. R. (1988) *The New S Language.* Wadsworth & Brooks/Cole.

See Also

`dput`, `dget`,`write`.

Examples

```
x <- 1; y <- 1:10
dump(ls(patt='^[xyz]'), "xyz.Rdmped")
unlink("xyz.Rdmped")
```

duplicated *Determine Duplicate Elements*

Description

Determines which elements of a vector of data frame are duplicates of elements with smaller subscripts, and returns a logical vector indicating which elements (rows) are duplicates.

Usage

```
duplicated(x, incomparables = FALSE, ...)

## S3 method for class 'array':
duplicated(x, incomparables = FALSE, MARGIN = 1, ...)
```

Arguments

`x`	an atomic vector or a data frame or an array.
`incomparables`	a vector of values that cannot be compared. Currently, `FALSE` is the only possible value, meaning that all values can be compared.
`...`	arguments for particular methods.
`MARGIN`	the array margin to be held fixed: see `apply`.

Details

This is a generic function with methods for vectors, data frames and arrays (including matrices).

The data frame method works by pasting together a character representation of the rows separated by
`r`, so may be imperfect if the data frame has characters with embedded carriage returns or columns which do not reliably map to characters.

The array method calculates for each element of the sub-array specified by `MARGIN` if the remaining dimensions are identical to those for an earlier element (in row-major order). This would most commonly be used to find duplicated rows (the default) or columns (with `MARGIN = 2`).

References

Becker, R. A., Chambers, J. M. and Wilks, A. R. (1988) *The New S Language.* Wadsworth & Brooks/Cole.

See Also

`unique`.

Examples

```
x <- c(9:20, 1:5, 3:7, 0:8)
## extract unique elements
(xu <- x[!duplicated(x)])
## xu == unique(x) but unique(x) is more efficient

data(iris)
duplicated(iris)[140:143]

data(iris3)
duplicated(iris3, MARGIN = c(1, 3))
```

`dyn.load` *Foreign Function Interface*

Description

Load or unload shared libraries, and test whether a C function or Fortran subroutine is available.

Usage

```
dyn.load(x, local = TRUE, now = TRUE)
dyn.unload(x)

is.loaded(symbol, PACKAGE="")
symbol.C(name)
symbol.For(name)
```

Arguments

`x`	a character string giving the pathname to a shared library or DLL.
`local`	a logical value controlling whether the symbols in the shared library are stored in their own local table and not shared across shared libraries, or added to the global symbol table. Whether this has any effect is system-dependent.
`now`	a logical controlling whether all symbols are resolved (and relocated) immediately the library is loaded or deferred until they are used. This control is useful for developers testing whether a library is complete and has all the necessary symbols and for users to ignore missing symbols. Whether this has any effect is system-dependent.
`symbol`	a character string giving a symbol name.
`PACKAGE`	if supplied, confine the search for the `name` to the DLL given by this argument (plus the conventional extension, `.so`, `.sl`, `.dll`, ...). This is intended to add safety for packages, which can ensure by using this argument that no other package can override their external symbols. Use `PACKAGE="base"` for symbols linked in to R. This is used in the same way as in `.C`, `.Call`, `.Fortran` and `.External` functions

`name`	a character string giving either the name of a C function or Fortran subroutine. Fortran names probably need to be given entirely in lower case (but this may be system-dependent).

Details

See 'See Also' and the *Writing R Extensions* manual for how to create a suitable shared library. Note that unlike some versions of S-PLUS, `dyn.load` does not load an object (`.o`) file but a shared library or DLL.

Unfortunately a very few platforms (Compaq Tru64) do not handle the `PACKAGE` argument correctly, and may incorrectly find symbols linked into R.

The additional arguments to `dyn.load` mirror the different aspects of the mode argument to the dlopen() routine on UNIX systems. They are available so that users can exercise greater control over the loading process for an individual library. In general, the defaults values are appropriate and one should override them only if there is good reason and you understand the implications.

The `local` argument allows one to control whether the symbols in the DLL being attached are visible to other DLLs. While maintaining the symbols in their own namespace is good practice, the ability to share symbols across related "chapters" is useful in many cases. Additionally, on certain platforms and versions of an operating system, certain libraries must have their symbols loaded globally to successfully resolve all symbols.

One should be careful of the potential side-effect of using lazy loading via the `now` argument as `FALSE`. If a routine is called that has a missing symbol, the process will terminate immediately and unsaved session variables will be lost. The intended use is for library developers to call specify a value `TRUE` to check that all symbols are actually resolved and for regular users to all with `FALSE` so that missing symbols can be ignored and the available ones can be called.

The initial motivation for adding these was to avoid such termination in the `_init()` routines of the Java virtual machine library. However, symbols loaded locally may not be (read probably) available to other DLLs. Those added to the global table are available to all other elements of the application and so can be shared across two different DLLs.

Some systems do not provide (explicit) support for local/global and lazy/eager symbol resolution. This can be the source of subtle bugs. One can arrange to have warning messages emitted when unsupported options are used. This is done by setting either of the options `verbose` or `warn` to be non-zero via the `options` function. Currently, we know of only 2 platforms that do not provide a value for local load (RTLD_LOCAL). These are IRIX6.4 and unpatched versions of Solaris 2.5.1.

There is a short discussion of these additional arguments with some example code available at `http://cm.bell-labs.com/stat/duncan/R/dynload`.

Value

The function `dyn.load` is used for its side effect which links the specified shared library to the executing R image. Calls to `.C`, `.Fortran` and `.External` can then be used to execute compiled C functions or Fortran subroutines contained in the library.

The function `dyn.unload` unlinks the shared library.

Functions `symbol.C` and `symbol.For` map function or subroutine names to the symbol name in the compiled code: `is.loaded` checks if the symbol name is loaded and hence available for use in `.C` or `.Fortran`.

Note

The creation of shared libraries and the runtime linking of them into executing programs is very platform dependent. In recent years there has been some simplification in the process because the C subroutine call `dlopen` has become the standard for doing this under UNIX. Under UNIX `dyn.load` uses the `dlopen` mechanism and should work on all platforms which support it. On Windows it uses the standard mechanisms for loading 32-bit DLLs.

The original code for loading DLLs in UNIX was provided by Heiner Schwarte. The compatibility code for HP-UX was provided by Luke Tierney.

References

Becker, R. A., Chambers, J. M. and Wilks, A. R. (1988) *The New S Language*. Wadsworth & Brooks/Cole.

See Also

`library.dynam` to be used inside a package's `.First.lib` initialization.

`SHLIB` for how to create suitable shared objects.

`.C`, `.Fortran`, `.External`, `.Call`.

Examples

```
# probably TRUE, as mva is loaded
is.loaded(symbol.For("hcass2"))
```

edit *Invoke a Text Editor*

Description

Invoke a text editor on an R object.

Usage

```
## Default S3 method:
edit(name = NULL, file = "", editor=getOption("editor"), ...)
vi(name = NULL, file = "")
emacs(name = NULL, file = "")
pico(name = NULL, file = "")
xemacs(name = NULL, file = "")
xedit(name = NULL, file = "")
```

Arguments

`name`	a named object that you want to edit. If name is missing then the file specified by `file` is opened for editing.
`file`	a string naming the file to write the edited version to.
`editor`	a string naming the text editor you want to use. On Unix the default is set from the environment variables `EDITOR` or `VISUAL` if either is set, otherwise `vi` is used. On Windows it defaults to `notepad`.
`...`	further arguments to be passed to or from methods.

Details

`edit` invokes the text editor specified by `editor` with the object `name` to be edited. It is a generic function, currently with a default method and one for data frames and matrices.

`data.entry` can be used to edit data, and is used by `edit` to edit matrices and data frames on systems for which `data.entry` is available.

It is important to realize that `edit` does not change the object called `name`. Instead, a copy of name is made and it is that copy which is changed. Should you want the changes to apply to the object `name` you must assign the result of `edit` to `name`. (Try `fix` if you want to make permanent changes to an object.)

In the form `edit(name)`, `edit` deparses `name` into a temporary file and invokes the editor `editor` on this file. Quitting from the editor causes `file` to be parsed and that value returned. Should an error occur in parsing, possibly due to incorrect syntax, no value is returned. Calling `edit()`, with no arguments, will result in the temporary file being reopened for further editing.

Note

The functions `vi`, `emacs`, `pico`, `xemacs`, `xedit` rely on the corresponding editor being available and being on the path. This is system-dependent.

See Also

`edit.data.frame`, `data.entry`, `fix`.

Examples

```
# use xedit on the function mean and assign the changes
mean <- edit(mean, editor = "xedit")

# use vi on mean and write the result to file mean.out
vi(mean, file = "mean.out")
```

edit.data.frame *Edit Data Frames and Matrices*

Description

Use data editor on data frame or matrix contents.

Usage

```
## S3 method for class 'data.frame':
edit(name, factor.mode = c("character", "numeric"),
  edit.row.names = any(row.names(name) != 1:nrow(name)),
  ...)

## S3 method for class 'matrix':
edit(name,
  edit.row.names = any(rownames(name) != 1:nrow(name)), ...)
```

Arguments

`name`	A data frame or matrix.
`factor.mode`	How to handle factors (as integers or using character levels) in a data frame.
`edit.row.names`	logical. Show the row names be displayed as a separate editable column?
`...`	further arguments passed to or from other methods.

Details

At present, this only works on simple data frames containing numeric, logical or character vectors and factors. Factors are represented in the spreadsheet as either numeric vectors (which is more suitable for data entry) or character vectors (better for browsing). After editing, vectors are padded with `NA` to have the same length and factor attributes are restored. The set of factor levels can not be changed by editing in numeric mode; invalid levels are changed to `NA` and a warning is issued. If new factor levels are introduced in character mode, they are added at the end of the list of levels in the order in which they encountered.

It is possible to use the data-editor's facilities to select the mode of columns to swap between numerical and factor columns in a data frame. Changing any column in a numerical matrix to character will cause the result to be coerced to a character matrix. Changing the mode of logical columns is not supported.

Value

The edited data frame.

Note

`fix(dataframe)` works for in-place editing by calling this function.

If the data editor is not available, a dump of the object is presented for editing using the default method of `edit`.

At present the data editor is limited to 65535 rows.

Author(s)

Peter Dalgaard

See Also

`data.entry`, `edit`

Examples

```
data(InsectSprays)
edit(InsectSprays)
edit(InsectSprays, factor.mode="numeric")
```

`environment` *Environment Access*

Description

Get, set, test for and create environments.

Usage

```
environment(fun = NULL)
environment(fun) <- value
is.environment(obj)
.GlobalEnv
globalenv()
new.env(hash=FALSE, parent=parent.frame())
parent.env(env)
parent.env(env) <- value
```

Arguments

`fun`	a `function`, a `formula`, or `NULL`, which is the default.
`value`	an environment to associate with the function
`obj`	an arbitrary R object.
`hash`	a logical, if `TRUE` the environment will be hashed
`parent`	an environment to be used as the parent of the environment created.
`env`	an environment

Details

The global environment `.GlobalEnv` is the first item on the search path, more often known as the user's workspace. It can also be accessed by `globalenv()`.

The variable `.BaseNamespaceEnv` is part of some experimental support for name space management.

The replacement function `parent.env<-` is extremely dangerous as it can be used to destructively change environments in ways that violate assumptions made by the internal C code. It may be removed in the near future.

`is.environment` is generic: you can write methods to handle specific classes of objects, see InternalMethods.

Value

If `fun` is a function or a formula then `environment(fun)` returns the environment associated with that function or formula. If `fun` is `NULL` then the current evaluation environment is returned.

The assignment form sets the environment of the function or formula `fun` to the `value` given.

`is.environment(obj)` returns `TRUE` if and only if `obj` is an `environment`.

`new.env` returns a new (empty) environment enclosed in the parent's environment, by default.

`parent.env` returns the parent environment of its argument.

`parent.env<-` sets the parent environment of its first argument.

See Also

The `envir` argument of `eval`.

Examples

```
## all three give the same:
environment()
environment(environment)
.GlobalEnv

ls(envir=environment(approxfun(1:2,1:2, method="const")))

is.environment(.GlobalEnv) # TRUE

e1 <- new.env(TRUE, NULL)
e2 <- new.env(FALSE, NULL)
assign("a", 3, env=e2)
parent.env(e1) <- e2
get("a", env=e1)
```

`eval` *Evaluate an (Unevaluated) Expression*

Description

Evaluate an R expression in a specified environment.

Usage

```
eval(expr, envir = parent.frame(), enclos =
 if(is.list(envir) || is.pairlist(envir)) parent.frame())
evalq(expr, envir, enclos)
eval.parent(expr, n = 1)
local(expr, envir = new.env())
```

Arguments

`expr`	object of mode `expression` or `call` or an "unevaluated expression".
`envir`	the `environment` in which `expr` is to be evaluated. May also be a list, a data frame, or an integer as in `sys.call`.
`enclos`	Relevant when `envir` is a list or a data frame. Specifies the enclosure, i.e., where R looks for objects not found in `envir`.
`n`	parent generations to go back

Details

`eval` evaluates the expression `expr` argument in the environment specified by `envir` and returns the computed value. If `envir` is not specified, then `sys.frame(sys.frame())`, the environment where the call to `eval` was made is used.

The `evalq` form is equivalent to `eval(quote(expr), ...)`.

As `eval` evaluates its first argument before passing it to the evaluator, it allows you to assign complicated expressions to symbols and then evaluate them. `evalq` avoids this.

`eval.parent(expr, n)` is a shorthand for `eval(expr, parent.frame(n))`.

`local` evaluates an expression in a local environment. It is equivalent to `evalq` except the its default argument creates a new, empty environment. This is useful to create anonymous recursive functions and as a kind of limited namespace feature since variables defined in the environment are not visible from the outside.

Note

Due to the difference in scoping rules, there are some differences between R and S in this area. In particular, the default enclosure in S is the global environment.

When evaluating expressions in data frames that have been passed as an argument to a function, the relevant enclosure is often the caller's environment, i.e., one needs `eval(x, data, parent.frame())`.

References

Becker, R. A., Chambers, J. M. and Wilks, A. R. (1988) *The New S Language*. Wadsworth & Brooks/Cole. (`eval` only.)

See Also

`expression`, `quote`, `sys.frame`, `parent.frame`, `environment`.

Examples

```
eval(2 ^ 2 ^ 3)
mEx <- expression(2^2^3); mEx; 1 + eval(mEx)
eval({ xx <- pi; xx^2}) ; xx

a <- 3 ; aa <- 4 ;
evalq(evalq(a+b+aa, list(a=1)), list(b=5)) # == 10
a <- 3 ; aa <- 4 ;
evalq(evalq(a+b+aa, -1), list(b=5))        # == 12

ev <- function() {
   e1 <- parent.frame()
   ## Evaluate a in e1
   aa <- eval(expression(a),e1)
   ## evaluate the expression bound to a in e1
   a <- expression(x+y)
   list(aa = aa, eval = eval(a, e1))
}
tst.ev <- function(a = 7) { x <- pi; y <- 1; ev() }
```

```
tst.ev() # aa : 7,  eval : 4.14

##
## Uses of local()
##

# Mutual recursives.
# gg gets value of last assignment, an anonymous version
# of f.
gg <- local({
    k <- function(y)f(y)
    f <- function(x) if(x) x*k(x-1) else 1
})
gg(10)
sapply(1:5, gg)

# Nesting locals. a is private storage accessible to k
gg <- local({
    k <- local({
        a <- 1
        function(y){print(a <<- a+1);f(y)}
    })
    f <- function(x) if(x) x*k(x-1) else 1
})
sapply(1:5, gg)

ls(envir=environment(gg))
ls(envir=environment(get("k", envir=environment(gg))))
```

example — *Run an Examples Section from the Online Help*

Description

Run all the R code from the **Examples** part of R's online help topic `topic` with two possible exceptions, `dontrun` and `dontshow`, see Details below.

Usage

```
example(topic, package = .packages(), lib.loc = NULL,
  local = FALSE, echo = TRUE, verbose = getOption("verbose"),
  prompt.echo = paste(abbreviate(topic, 6),"> ", sep=""))
```

Arguments

`topic`	name or literal character string: the online `help` topic the examples of which should be run.
`package`	a character vector with package names. By default, all packages in the search path are used.
`lib.loc`	a character vector of directory names of R libraries, or `NULL`. The default value of `NULL` corresponds to all libraries currently known. If the default is used, the loaded packages are searched before the libraries.
`local`	logical: if `TRUE` evaluate locally, if `FALSE` evaluate in the workspace.
`echo`	logical; if `TRUE`, show the R input when sourcing.
`verbose`	logical; if `TRUE`, show even more when running example code.
`prompt.echo`	character; gives the prompt to be used if `echo = TRUE`.

Details

If `lib.loc` is not specified, the packages are searched for amongst those already loaded, then in the specified libraries. If `lib.loc` is specified, they are searched for only in the specified libraries, even if they are already loaded from another library.

An attempt is made to load the package before running the examples, but this will not replace a package loaded from another location.

If `local=TRUE` objects are not created in the workspace and so not available for examination after `example` completes: on the other hand they cannot clobber objects of the same name in the workspace.

As detailed in the manual *Writing* R *Extensions*, the author of the help page can markup parts of the examples for two exception rules

`dontrun` encloses code that should not be run.

`dontshow` encloses code that is invisible on help pages, but will be run both by the package checking tools, and the `example()` function. This was previously `testonly`, and that form is still accepted.

Value

(the value of the last evaluated expression).

Note

The examples can be many small files. On some file systems it is desirable to save space, and the files in the '`R-ex`' directory of an installed package can be zipped up as a zip archive '`Rex.zip`'.

Author(s)

Martin Maechler and others

See Also

`demo`

Examples

```
example(InsectSprays)
## force use of the standard package 'eda':
example("smooth", package="eda", lib.loc=.Library)
```

`exists` *Is an Object Defined?*

Description

Search for an R object of the given name on the search path.

Usage

```
exists(x, where = -1, envir = , frame, mode = "any",
       inherits = TRUE)
```

Arguments

`x`	a variable name (given as a character string).
`where`	where to look for the object (see the details section); if omitted, the function will search, as if the name of the object appeared unquoted in an expression.
`envir`	an alternative way to specify an environment to look in, but it's usually simpler to just use the `where` argument.
`frame`	a frame in the calling list. Equivalent to giving `where` as `sys.frame(frame)`.
`mode`	the mode of object sought.
`inherits`	should the enclosing frames of the environment be inspected.

Details

The `where` argument can specify the environment in which to look for the object in any of several ways: as an integer (the position in the `search` list); as the character string name of an element in the search list; or as an `environment` (including using `sys.frame` to access the currently active function calls). The `envir` argument is an alternative way to specify an environment, but is primarily there for back compatibility.

This function looks to see if the name `x` has a value bound to it. If `inherits` is `TRUE` and a value is not found for `x`, then the parent frames of the environment are searched until the name `x` is encountered. **Warning:** This is the default behaviour for R but not for S.

If `mode` is specified then only objects of that mode are sought. The function returns `TRUE` if the variable is encountered and `FALSE` if not.

The `mode` includes collections such as `"numeric"` and `"function"`: any member of the collection will suffice.

Value

Logical, true if and only if the object is found on the search path.

References

Becker, R. A., Chambers, J. M. and Wilks, A. R. (1988) *The New S Language.* Wadsworth & Brooks/Cole.

See Also

`get`.

Examples

```
## Define a substitute function if necessary:
if(!exists("some.fun", mode="function"))
 some.fun <- function(x) { cat("some.fun(x)\n"); x }
search()
exists("ls", 2) # true even though ls is in pos=3
exists("ls", 2, inherits=FALSE) # false
```

`expression` *Unevaluated Expressions*

Description

Creates or tests for objects of mode `"expression"`.

Usage

```
expression(...)

is.expression(x)
as.expression(x, ...)
```

Arguments

`...`	valid R expressions.
`x`	an arbitrary R object.

Value

`expression` returns a vector of mode `"expression"` containing its arguments as unevaluated "calls".

`is.expression` returns `TRUE` if `expr` is an expression object and `FALSE` otherwise.

`as.expression` attempts to coerce its argument into an expression object.

References

Becker, R. A., Chambers, J. M. and Wilks, A. R. (1988) *The New S Language.* Wadsworth & Brooks/Cole.

See Also

`call`, `eval`, `function`. Further, `text` and `legend` for plotting math expressions.

Examples

```
length(ex1 <- expression(1+ 0:9)) # 1
ex1
eval(ex1) # 1:10

length(ex3 <- expression(u,v, 1+ 0:9)) # 3
mode(ex3 [3]) # expression
mode(ex3[[3]]) # call
rm(ex3)
```

Extract *Extract or Replace Parts of an Object*

Description

Operators act on vectors, arrays and lists to extract or replace subsets.

Usage

```
x[i]
x[i, j, ... , drop=TRUE]
x[[i]]
x[[i, j, ...]]
x$name

.subset(x, ...)
.subset2(x, ...)
```

Arguments

`x` object from which to extract elements or in which to replace elements.

`i, j, ..., name`
elements to extract or replace. `i, j` are `numeric` or `character` or empty whereas `name` must be character or an (unquoted) name. Numeric values are coerced to integer as by `as.integer`.

For `[`-indexing only: `i, j, ...` can be logical vectors, indicating elements/slices to select. Such vectors are recycled if necessary to match the corresponding extent. When indexing arrays, `i` can be a (single) matrix with as many columns as there are dimensions of `x`; the result is then a vector with elements corresponding to the sets of indices in each row of `i`.

`drop` For matrices, and arrays. If `TRUE` the result is coerced to the lowest possible dimension (see examples below). This only works for extracting elements, not for the replacement forms.

Details

These operators are generic. You can write methods to handle subsetting of specific classes of objects, see InternalMethods as well as `[`.

`data.frame` and `[.factor`. The descriptions here apply only to the default methods.

The most important distinction between `[`, `[[` and `$` is that the `[` can select more than one element whereas the other two select a single element. `$` does not allow computed indices, whereas `[[` does. `x$name` is equivalent to `x[["name"]]` if `x` is recursive (see `is.recursive`) and `NULL` otherwise.

The `[[` operator requires all relevant subscripts to be supplied. With the `[` operator an empty index (a comma separated blank) indicates that all entries in that dimension are selected.

If one of these expressions appears on the left side of an assignment then that part of `x` is set to the value of the right hand side of the assignment.

Indexing by factors is allowed and is equivalent to indexing by the numeric codes (see `factor`) and not by the character values which are printed (for which use `[as.character(i)]`).

When operating on a list, the `[[` operator gives the specified element of the list while the `[` operator returns a list with the specified element(s) in it.

As from R 1.7.0 `[[` can be applied recursively to lists, so that if the single index `i` is a vector of length `p`, `alist[[i]]` is equivalent to `alist[[i1]]...[[ip]]` providing all but the final indexing results in a list.

The operators `$` and `$<-` do not evaluate their second argument. It is translated to a string and that string is used to locate the correct component of the first argument.

When `$<-` is applied to a `NULL` `x`, it coerces `x` to `list()`. This is what happens with `[[<-` is `y` is of length greater than one: if `y` has length 1 or 0, `x` is coerced to a zero-length vector of the type of `value`,

The functions `.subset` and `.subset2` are essentially equivalent to the `[` and `[[` operators, except that methods dispatch does not take place. This is to avoid expensive unclassing when applying the default method to an object. They should not normally be invoked by end users.

References

Becker, R. A., Chambers, J. M. and Wilks, A. R. (1988) *The New S Language*. Wadsworth & Brooks/Cole.

See Also

`list`, `array`, `matrix`.

`[.data.frame` and `[.factor` for the behaviour when applied to data.frame and factors.

`Syntax` for operator precedence, and the *R Language* reference manual about indexing details.

Examples

```
x <- 1:12; m <- matrix(1:6,nr=2);
li <- list(pi=pi, e = exp(1))
x[10]                    # the tenth element of x
m[1,]                    # the first row of matrix m
m[1, , drop = FALSE]     # is a 1-row matrix
m[,c(TRUE,FALSE,TRUE)] # logical indexing
m[cbind(c(1,2,1),3:1)] # matrix index
li[[1]]                  # the first element of list li
y <- list(1,2,a=4,5)
y[c(3,4)] # a list containing elements 3 and 4 of y
y$a       # the element of y named a

## non-integer indices are truncated:
(i <- 3.999999999) # "4" is printed
(1:5)[i]  # 3

## recursive indexing into lists
z <- list( a=list( b=9, c='hello'), d=1:5)
unlist(z)
z[[c(1, 2)]]
z[[c(1, 2, 1)]]  # both "hello"
z[[c("a", "b")]] <- "new"
unlist(z)
```

`Extract.data.frame` *Extract or Replace Parts of a Data Frame*

Description

Extract or replace subsets of data frames.

Usage

```
x[i]
x[i] <- value
x[i, j, drop = TRUE]
x[i, j] <- value

x[[i]]
x[[i]] <- value
x[[i, j]]
x[[i, j]] <- value

x$name
x$name <- value
```

Arguments

`x`	data frame.
`i, j`	elements to extract or replace. `i, j` are `numeric` or `character` or, for `[` only, empty. Numeric values are coerced to integer as if by `as.integer`. For replacement by `[`, a logical matrix is allowed.
`drop`	logical. If `TRUE` the result is coerced to the lowest possible dimension: however, see the Warning below.
`value`	A suitable replacement value: it will be repeated a whole number of times if necessary and it may be coerced: see the Coercion section. If `NULL`, deletes the column if a single column is selected.
`name`	name or literal character string.

Details

Data frames can be indexed in several modes. When `[` and `[[` are used with a single index, they index the data frame as if it were a list. In this usage a `drop` argument is ignored, with a warning. Using `$` is equivalent to using `[[` with a single index.

When `[` and `[[` are used with two indices they act like indexing a matrix: `[[` can only be used to select one element.

If `[` returns a data frame it will have unique (and non-missing) row names, if necessary transforming the row names using `make.unique`. Similarly, column names will be transformed (if columns are selected more than once).

When `drop =TRUE`, this is applied to the subsetting of any matrices contained in the data frame as well as to the data frame itself.

The replacement methods can be used to add whole column(s) by specifying non-existent column(s), in which case the column(s) are added at the right-hand edge of the data frame and numerical indices must be contiguous to existing indices. On the other hand, rows can be added at any row after the current last row, and the columns will be in-filled with missing values.

For `[` the replacement value can be a list: each element of the list is used to replace (part of) one column, recycling the list as necessary. If the columns specified by number are created, the names (if any) of the corresponding list elements are used to name the columns. If the replacement is not selecting rows, list values can contain `NULL` elements which will cause the corresponding columns to be deleted.

Matrixing indexing using `[` is not recommended, and barely supported. For extraction, `x` is first coerced to a matrix. For replacement a logical matrix (only) can be used to select the elements to be replaced in the same ways as for a matrix. Missing values in the matrix are treated as false, unlike S which does not replace them but uses up the corresponding values in `value`.

Value

For `[` a data frame, list or a single column (the latter two only when dimensions have been dropped). If matrix indexing is used for extraction a matrix results.

For `[[` a column of the data frame (extraction with one index) or a length-one vector (extraction with two indices).

For `[<-`, `[[<-` and `$<-`, a data frame.

Coercion

The story over when replacement values are coerced is a complicated one, and one that has changed during R's development. This section is a guide only.

When `[` and `[[` are used to add or replace a whole column, no coercion takes place but `value` will be replicated (by calling the generic function `rep`) to the right length if an exact number of repeats can be used.

When `[` is used with a logical matrix, each value is coerced to the type of the column in which it is to be placed.

When `[` and `[[` are used with two indices, the column will be coerced as necessary to accommodate the value.

Warning

Although the default for `drop` is `TRUE`, the default behaviour when only one *row* is left is equivalent to specifying `drop = FALSE`. To drop from a data frame to a list, `drop = FALSE` has to specified explicitly.

See Also

`subset` which is often easier for extraction, `data.frame`, `Extract`.

Examples

```
data(swiss)
sw <- swiss[1:5, 1:4]  # select a manageable subset

sw[1:3]       # select columns
sw[, 1:3]     # same
sw[4:5, 1:3] # select rows and columns
sw[1]         # a one-column data frame
sw[, 1, drop = FALSE]  # the same
sw[, 1]       # a (unnamed) vector
sw[[1]]       # the same

sw[1,]        # a one-row data frame
sw[1,, drop=TRUE]  # a list

# duplicate row, unique row names are created
swiss[ c(1, 1:2), ]

sw[sw <= 6] <- 6  # logical matrix indexing
```

```
sw

## adding a column
sw["new1"] <- LETTERS[1:5]   # adds a character column
sw[["new2"]] <- letters[1:5] # ditto
sw[, "new3"] <- LETTERS[1:5] # ditto, but this got
                             # converted to a factor in 1.7.x
sw$new4 <- 1:5
sapply(sw, class)
sw$new4 <- NULL              # delete the column
sw
# delete col7, update 6, append
sw[6:8] <- list(letters[10:14], NULL, aa=1:5)
sw

## matrices in a data frame
A <- data.frame(x=1:3, y=I(matrix(4:6)),
                z=I(matrix(letters[1:9],3,3)))
A[1:3, "y"] # a matrix, was a vector prior to 1.8.0
A[1:3, "z"] # a matrix
A[, "y"]    # a matrix
```

Extract.factor *Extract or Replace Parts of a Factor*

Description

Extract or replace subsets of factors.

Usage

```
x[i, drop = FALSE]

x[i] <- value
```

Arguments

`x`	a factor
`i`	a specification of indices – see `Extract`.
`drop`	logical. If true, unused levels are dropped.
`value`	character: a set of levels. Factor values are coerced to character.

Details

When unused levels are dropped the ordering of the remaining levels is preserved.

If `value` is not in `levels(x)`, a missing value is assigned with a warning.

Any `contrasts` assigned to the factor are preserved unless `drop=TRUE`.

Value

A factor with the same set of levels as `x` unless `drop=TRUE`.

See Also

`factor`, `Extract`.

Examples

```
## following example(factor)
(ff <- factor(substring("statistics", 1:10, 1:10),
              levels=letters))
ff[, drop=TRUE]
factor(letters[7:10])[2:3, drop = TRUE]
```

factor *Factors*

Description

The function `factor` is used to encode a vector as a factor (the names category and enumerated type are also used for factors). If `ordered` is `TRUE`, the factor levels are assumed to be ordered. For compatibility with S there is also a function `ordered`.

`is.factor`, `is.ordered`, `as.factor` and `as.ordered` are the membership and coercion functions for these classes.

Usage

```
factor(x, levels = sort(unique.default(x), na.last = TRUE),
  labels = levels, exclude = NA, ordered = is.ordered(x))
ordered(x, ...)

is.factor(x)
is.ordered(x)

as.factor(x)
as.ordered(x)
```

Arguments

`x`	a vector of data, usually taking a small number of distinct values
`levels`	an optional vector of the values that `x` might have taken. The default is the set of values taken by `x`, sorted into increasing order.
`labels`	*either* an optional vector of labels for the levels (in the same order as `levels` after removing those in `exclude`), *or* a character string of length 1.
`exclude`	a vector of values to be excluded when forming the set of levels. This should be of the same type as `x`, and will be coerced if necessary.
`ordered`	logical flag to determine if the levels should be regarded as ordered (in the order given).
`...`	(in `ordered(.)`): any of the above, apart from `ordered` itself.

Details

The type of the vector `x` is not restricted.

Ordered factors differ from factors only in their class, but methods and the model-fitting functions treat the two classes quite differently.

The encoding of the vector happens as follows. First all the values in `exclude` are removed from `levels`. If `x[i]` equals `levels[j]`, then the `i`-th element of the result is `j`. If no match is found for `x[i]` in `levels`, then the `i`-th element of the result is set to `NA`.

Normally the 'levels' used as an attribute of the result are the reduced set of levels after removing those in `exclude`, but this can be altered by supplying `labels`. This should either be a set of new labels for the levels, or a character string, in which case the levels are that character string with a sequence number appended.

`factor(x, exclude=NULL)` applied to a factor is a no-operation unless there are unused levels: in that case, a factor with the reduced level set is returned. If `exclude` is used it should also be a factor with the same level set as `x` or a set of codes for the levels to be excluded.

The codes of a factor may contain `NA`. For a numeric `x`, set `exclude=NULL` to make `NA` an extra level (`"NA"`), by default the last level.

If `"NA"` is a level, the way to set a code to be missing is to use `is.na` on the left-hand-side of an assignment. Under those circumstances missing values are printed as `<NA>`.

`is.factor` is generic: you can write methods to handle specific classes of objects, see InternalMethods.

Value

`factor` returns an object of class `"factor"` which has a set of numeric codes the length of `x` with a `"levels"` attribute of mode `character`. If `ordered` is true (or `ordered` is used) the result has class `c("ordered", "factor")`.

Applying `factor` to an ordered or unordered factor returns a factor (of the same type) with just the levels which occur: see also `[.factor` for a more transparent way to achieve this.

`is.factor` returns `TRUE` or `FALSE` depending on whether its argument is of type factor or not. Correspondingly, `is.ordered` returns `TRUE` when its argument is ordered and `FALSE` otherwise.

`as.factor` coerces its argument to a factor. It is an abbreviated form of `factor`.

`as.ordered(x)` returns `x` if this is ordered, and `ordered(x)` otherwise.

Warning

The interpretation of a factor depends on both the codes and the `"levels"` attribute. Be careful only to compare factors with the same set of levels (in the same order). In particular, `as.numeric` applied to a factor is meaningless, and may happen by implicit coercion. To "revert" a factor `f` to its original numeric values, `as.numeric(levels(f))[f]` is recommended and slightly more efficient than `as.numeric(as.character(f))`.

The levels of a factor are by default sorted, but the sort order may well depend on the locale at the time of creation, and should not be assumed to be ASCII.

References

Chambers, J. M. and Hastie, T. J. (1992) *Statistical Models in S.* Wadsworth & Brooks/Cole.

See Also

`[.factor` for subsetting of factors.

`gl` for construction of "balanced" factors and `C` for factors with specified contrasts. `levels` and `nlevels` for accessing the levels, and `codes` to get integer codes.

Examples

```
(ff <- factor(substring("statistics", 1:10, 1:10),
  levels=letters))
as.integer(ff) # the internal codes
factor(ff)      # drops the levels that do not occur
ff[, drop=TRUE] # the same, more transparently

factor(letters[1:20], label="letter")

class(ordered(4:1)) # "ordered", inheriting from "factor"

## suppose you want "NA" as a level, and to allowing
## missing values.
(x <- factor(c(1, 2, "NA"), exclude = ""))
is.na(x)[2] <- TRUE
x  # [1] 1    <NA> NA, <NA> used because NA is a level.
is.na(x)
# [1] FALSE  TRUE FALSE
```

`file.access` *Ascertain File Accessibility*

Description

Utility function to access information about files on the user's file systems.

Usage

```
file.access(names, mode = 0)
```

Arguments

`names`	character vector containing file names.
`mode`	integer specifying access mode required.

Details

The `mode` value can be the exclusive or of the following values

0 test for existence.

1 test for execute permission.

2 test for write permission.

4 test for read permission.

Permission will be computed for real user ID and real group ID (rather than the effective IDs).

Value

An integer vector with values `0` for success and `-1` for failure.

Note

This is intended as a replacement for the S-PLUS function `access`, a wrapper for the C function of the same name, which explains the return value encoding. Note that the return value is **false** for **success**.

See Also

`file.info`

Examples

```
fa <- file.access(dir("."))
table(fa) # count successes & failures
```

`file.choose` *Choose a File Interactively*

Description

Choose a file interactively.

Usage

```
file.choose(new = FALSE)
```

Arguments

`new`	Logical: choose the style of dialog box presented to the user: at present only new = FALSE is used.

Value

A character vector of length one giving the file path.

`file.info` *Extract File Information*

Description

Utility function to extract information about files on the user's file systems.

Usage

```
file.info(...)
```

Arguments

`...`	character vectors containing file names.

Details

What is meant by "file access" and hence the last access time is system-dependent.

On most systems symbolic links are followed, so information is given about the file to which the link points rather than about the link.

Value

A data frame with row names the file names and columns

`size`	integer: File size in bytes.
`isdir`	logical: Is the file a directory?
`mode`	integer of class `"octmode"`. The file permissions, printed in octal, for example `644`.
`mtime, ctime, atime`	integer of class `"POSIXct"`: file modification, creation and last access times.
`uid`	integer: the user ID of the file's owner.
`gid`	integer: the group ID of the file's group.
`uname`	character: `uid` interpreted as a user name.
`grname`	character: `gid` interpreted as a group name.

Unknown user and group names will be `NA`.

Entries for non-existent or non-readable files will be `NA`. The `uid`, `gid`, `uname` and `grname` columns may not be supplied on a non-POSIX Unix system.

Note

This function will only be operational on systems with the `stat` system call, but that seems very widely available.

See Also

`files`, `file.access`, `list.files`, and `DateTimeClasses` for the date formats.

Examples

```
ncol(finf <- file.info(dir())) # at least six
finf # the whole list
## Those that are more than 100 days old :
finf[
  difftime(Sys.time(),finf[,"mtime"],units="days")>100, 1:4
]

file.info("no-such-file-exists")
```

`file.path` *Construct Path to File*

Description

Construct the path to a file from components in a platform-independent way.

Usage

```
file.path(..., fsep = .Platform$file.sep)
```

Arguments

`...`	character vectors.
`fsep`	the path separator to use.

Value

A character vector of the arguments concatenated term-by-term and separated by `fsep` if all arguments have positive length; otherwise, an empty character vector.

`file.show` *Display One or More Files*

Description

Display one or more files.

Usage

```
file.show(..., header = rep("",nfiles),
          title = "R Information",
          delete.file=FALSE, pager=getOption("pager"))
```

Arguments

`...`	one or more character vectors containing the names of the files to be displayed.
`header`	character vector (of the same length as the number of files specified in `...`) giving a header for each file being displayed. Defaults to empty strings.
`title`	an overall title for the display. If a single separate window is used for the display, `title` will be used as the window title. If multiple windows are used, their titles should combine the title and the file-specific header.
`delete.file`	should the files be deleted after display? Used for temporary files.
`pager`	the pager to be used.

Details

This function provides the core of the R help system, but it can be used for other purposes as well.

Note

How the pager is implemented is highly system dependent.

The basic Unix version concatenates the files (using the headers) to a temporary file, and displays it in the pager selected by the `pager` argument, which is a character vector specifying a system command to run on the set of files.

Most GUI systems will use a separate pager window for each file, and let the user leave it up while R continues running. The selection of such pagers could either be done using "magic" pager names being intercepted by lower-level code (such as "internal" and "console" on Windows), or by letting pager be an R function which will be called with the same arguments as file.show and take care of interfacing to the GUI.

Not all implementations will honour delete.file.

Author(s)

Ross Ihaka, Brian Ripley.

See Also

files, list.files, help.

Examples

```
file.show(file.path(R.home(), "COPYRIGHTS"))
```

files *File Manipulation*

Description

These functions provide a low-level interface to the computer's file system.

Usage

```
file.create(...)
file.exists(...)
file.remove(...)
file.rename(from, to)
file.append(file1, file2)
file.copy(from, to, overwrite = FALSE)
file.symlink(from, to)
dir.create(path)
```

Arguments

`..., file1, file2, from, to`	character vectors, containing file names.
`path`	a character vector containing a single path name.
`overwrite`	logical; should the destination files be overwritten?

Details

The `...` arguments are concatenated to form one character string: you can specify the files separately or as one vector. All of these functions expand path names: see `path.expand`.

`file.create` creates files with the given names if they do not already exist and truncates them if they do.

`file.exists` returns a logical vector indicating whether the files named by its argument exist.

`file.remove` attempts to remove the files named in its argument.

`file.rename` attempts to rename a single file.

`file.append` attempts to append the files named by its second argument to those named by its first. The R subscript recycling rule is used to align names given in vectors of different lengths.

`file.copy` works in a similar way to `file.append` but with the arguments in the natural order for copying. Copying to existing destination files is skipped unless `overwrite = TRUE`. The `to` argument can specify a single existing directory.

`file.symlink` makes symbolic links on those Unix-like platforms which support them. The `to` argument can specify a single existing directory.

`dir.create` creates the last element of the path.

Value

`dir.create` and `file.rename` return a logical, true for success.

The remaining functions return a logical vector indicating which operation succeeded for each of the files attempted.

Author(s)

Ross Ihaka, Brian Ripley

See Also

`file.info`, `file.access`, `file.path`, `file.show`, `list.files`, `unlink`, `basename`, `path.expand`.

Examples

```
cat("file A\n", file="A")
cat("file B\n", file="B")
file.append("A", "B")
file.create("A")
file.append("A", rep("B", 10))
if(interactive()) file.show("A")
file.copy("A", "C")
dir.create("tmp")
file.copy(c("A", "B"), "tmp")
list.files("tmp")
setwd("tmp")
file.remove("B")
file.symlink(file.path("..", c("A", "B")), ".")
setwd("..")
unlink("tmp", recursive=TRUE)
file.remove("A", "B", "C")
```

`fivenum` *Tukey Five-Number Summaries*

Description

Returns Tukey's five number summary (minimum, lower-hinge, median, upper-hinge, maximum) for the input data.

Usage

```
fivenum(x, na.rm = TRUE)
```

Arguments

`x`	numeric, maybe including `NA`s and $\pm$`Inf`s.
`na.rm`	logical; if `TRUE`, all `NA` and `NaN`s are dropped, before the statistics are computed.

Value

A numeric vector of length 5 containing the summary information. See `boxplot.stats` for more details.

See Also

`IQR`, `boxplot.stats`, `median`, `quantile`, `range`.

Examples

```
fivenum(c(rnorm(100),-1:1/0))
```

fix *Fix an Object*

Description

`fix` invokes `edit` on `x` and then assigns the new (edited) version of `x` in the user's workspace.

Usage

```
fix(x, ...)
```

Arguments

`x`	the name of an R object, as a name or a character string.
`...`	arguments to pass to editor: see `edit`.

Details

The name supplied as `x` need not exist as an R object, when a function with no arguments and an empty body is supplied for editing.

See Also

`edit`, `edit.data.frame`

Examples

```
## Assume 'my.fun' is a user defined function :
fix(my.fun)
## now my.fun is changed
## Also,
fix(my.data.frame) # calls up data editor
fix(my.data.frame, factor.mode="char") # use of ...
```

force *Force evaluation of an Argument*

Description

Forces the evaluation of a function argument.

Usage

```
force(x)
```

Arguments

x	a formal argument.

Details

`force` forces the evaluation of a formal argument. This can be useful if the argument will be captured in a closure by the lexical scoping rules and will later be altered by an explicit assignment or an implicit assignment in a loop or an apply function.

Examples

```
f <- function(y) function() y
lf <- vector("list", 5)
for (i in seq(along = lf)) lf[[i]] <- f(i)
lf[[1]]()  # returns 5

g <- function(y) { force(y); function() y }
lg <- vector("list", 5)
for (i in seq(along = lg)) lg[[i]] <- g(i)
lg[[1]]()  # returns 1
```

Foreign *Foreign Function Interface*

Description

Functions to make calls to compiled code that has been loaded into R.

Usage

```
                .C(name, ..., NAOK=FALSE, DUP=TRUE, PACKAGE)
          .Fortran(name, ..., NAOK=FALSE, DUP=TRUE, PACKAGE)
         .External(name, ..., PACKAGE)
             .Call(name, ..., PACKAGE)
.External.graphics(name, ..., PACKAGE)
    .Call.graphics(name, ..., PACKAGE)
```

Arguments

`name`	a character string giving the name of a C function or Fortran subroutine.
`...`	arguments to be passed to the foreign function.
`NAOK`	if `TRUE` then any `NA` or `NaN` or `Inf` values in the arguments are passed on to the foreign function. If `FALSE`, the presence of `NA` or `NaN` or `Inf` values is regarded as an error.
`DUP`	if `TRUE` then arguments are "duplicated" before their address is passed to C or Fortran.
`PACKAGE`	if supplied, confine the search for the `name` to the DLL given by this argument (plus the conventional extension, `.so`, `.sl`, `.dll`, ...). This is intended to add safety for packages, which can ensure by using this argument that no other package can override their external symbols. Use `PACKAGE="base"` for symbols linked in to R.

Details

The functions `.C` and `.Fortran` can be used to make calls to C and Fortran code.

`.External` and `.External.graphics` can be used to call compiled code that uses R objects in the same way as internal R functions.

.Call and .Call.graphics can be used to call compiled code which makes use of internal R objects. The arguments are passed to the C code as a sequence of R objects. It is included to provide compatibility with S version 4.

For details about how to write code to use with .Call and .External, see the chapter on “System and foreign language interfaces” in “Writing R Extensions” in the ‘doc/manual’ subdirectory of the R source tree.

Value

The functions .C and .Fortran return a list similar to the ... list of arguments passed in, but reflecting any changes made by the C or Fortran code.

.External, .Call, .External.graphics, and .Call.graphics return an R object.

These calls are typically made in conjunction with dyn.load which links DLLs to R.

The .graphics versions of .Call and .External are used when calling code which makes low-level graphics calls. They take additional steps to ensure that the device driver display lists are updated correctly.

Argument types

The mapping of the types of R arguments to C or Fortran arguments in .C or .Fortran is

R	C	Fortran
integer	int *	integer
numeric	double *	double precision
– or –	float *	real
complex	Rcomplex *	double complex
logical	int *	integer
character	char **	[see below]
list	SEXP *	not allowed
other	SEXP	not allowed

Numeric vectors in R will be passed as type double * to C (and as double precision to Fortran) unless (i) .C or .Fortran is used, (ii) DUP is false and (iii) the argument has attribute Csingle set to TRUE (use as.single or single). This mechanism is only intended to be used to facilitate the interfacing of existing C and Fortran code.

The C type Rcomplex is defined in ‘Complex.h’ as a typedef struct {double r; double i;}. Fortran type double complex is an extension to the Fortran standard, and the availability of a mapping of complex to Fortran may be compiler dependent.

Note: The C types corresponding to integer and logical are int, not long as in S.

The first character string of a character vector is passed as a C character array to Fortran: that string may be usable as character*255 if its true length is passed separately. Only up to 255 characters of the string are passed back. (How well this works, or even if it works at all, depends on the C and Fortran compilers and the platform.)

Missing (NA) string values are passed to .C as the string "NA". As the C char type can represent all possible bit patterns there appears to be no way to distinguish missing strings from the string "NA". If this distinction is important use .Call.

Functions, expressions, environments and other language elements are passed as the internal R pointer type SEXP. This type is defined in 'Rinternals.h' or the arguments can be declared as generic pointers, void *. Lists are passed as C arrays of SEXP and can be declared as void * or SEXP *. Note that you cannot assign values to the elements of the list within the C routine. Assigning values to elements of the array corresponding to the list bypasses R's memory management/garbage collection and will cause problems. Essentially, the array corresponding to the list is read-only. If you need to return S objects created within the C routine, use the .Call interface.

R functions can be invoked using call_S or call_R and can be passed lists or the simple types as arguments.

Header files for external code

Writing code for use with .External and .Call will use internal R structures. If possible use just those defined in 'Rinternals.h' and/or the macros in 'Rdefines.h', as other header files are not installed and are even more likely to be changed.

Note

DUP=FALSE is dangerous.

There are two dangers with using DUP=FALSE.

The first is that if you pass a local variable to .C/.Fortran with DUP=FALSE, your compiled code can alter the local variable and not just the copy in the return list. Worse, if you pass a local variable that is a formal parameter of the calling function, you may be able to change not only the local variable but the variable one level up. This will be very hard to trace.

The second is that lists are passed as a single R `SEXP` with `DUP=FALSE`, not as an array of `SEXP`. This means the accessor macros in '`Rinternals.h`' are needed to get at the list elements and the lists cannot be passed to `call_S`/`call_R`. New code using R objects should be written using `.Call` or `.External`, so this is now only a minor issue.

(Prior to R version 1.2.0 there has a third danger, that objects could be moved in memory by the garbage collector. The current garbage collector never moves objects.)

It is safe and useful to set `DUP=FALSE` if you do not change any of the variables that might be affected, e.g.,

`.C("Cfunction", input=x, output=numeric(10))`.

In this case the output variable did not exist before the call so it cannot cause trouble. If the input variable is not changed in the C code of `Cfunction` you are safe.

Neither `.Call` nor `.External` copy their arguments. You should treat arguments you receive through these interfaces as read-only.

References

Becker, R. A., Chambers, J. M. and Wilks, A. R. (1988) *The New S Language.* Wadsworth & Brooks/Cole. (`.C` and `.Fortran`.)

Chambers, J. M. (1998) *Programming with Data. A Guide to the S Language.* Springer. (`.Call`.)

See Also

`dyn.load`.

formals *Access to and Manipulation of the Formal Arguments*

Description

Get or set the formal arguments of a function.

Usage

```
formals(fun = sys.function(sys.parent()))
formals(fun, envir = parent.frame()) <- value
```

Arguments

`fun`	a function object, or see Details.
`envir`	environment in which the function should be defined.
`value`	a list of R expressions.

Details

For the first form, `fun` can be a character string naming the function to be manipulated, which is searched for from the parent environment. If it is not specified, the function calling `formals` is used.

Value

`formals` returns the formal argument list of the function specified.

The assignment form sets the formals of a function to the list on the right hand side.

See Also

`args` for a "human-readable" version, `alist`, `body`, `function`.

Examples

```
length(formals(lm))      # the number of formal arguments
names(formals(boxplot)) # formal arguments names

f <- function(x)a+b
formals(f) <- alist(a=,b=3) # function(a,b=3)a+b
f(2) # result = 5
```

format *Encode in a Common Format*

Description

Format an R object for pretty printing: `format.pval` is intended for formatting p-values.

Usage

```
format(x, ...)

## S3 method for class 'AsIs':
format(x, width = 12, ...)

## S3 method for class 'data.frame':
format(x, ..., justify = "none")

## Default S3 method:
format(x, trim = FALSE, digits = NULL,
       nsmall = 0, justify = c("left", "right", "none"),
       big.mark = "",   big.interval = 3,
     small.mark = "", small.interval = 5,
   decimal.mark = ".", ...)

## S3 method for class 'factor':
format(x, ...)

format.pval(pv, digits = max(1, getOption("digits") - 2),
            eps = .Machine$double.eps, na.form = "NA")

prettyNum(x, big.mark = "",   big.interval = 3,
           small.mark = "", small.interval = 5,
         decimal.mark = ".", ...)
```

Arguments

`x` any R object (conceptually); typically numeric.

`trim` logical; if `TRUE`, leading blanks are trimmed off the strings.

`digits` how many significant digits are to be used for `numeric` `x`. The default, `NULL`, uses `options()$digits`. This

	is a suggestion: enough decimal places will be used so that the smallest (in magnitude) number has this many significant digits.
`nsmall`	number of digits which will always appear to the right of the decimal point in formatting real/complex numbers in non-scientific formats. Allowed values `0 <= nsmall <= 20`.
`justify`	should character vector be left-justified, right-justified or left alone. When justifying, the field width is that of the longest string.
`big.mark`	character; if not empty used as mark between every `big.interval` decimals *before* (hence `big`) the decimal point.
`big.interval`	see `big.mark` above; defaults to 3.
`small.mark`	character; if not empty used as mark between every `small.interval` decimals *after* (hence `small`) the decimal point.
`small.interval`	see `small.mark` above; defaults to 5.
`decimal.mark`	the character used to indicate the numeric decimal point.
`pv`	a numeric vector.
`eps`	a numerical tolerance: see Details.
`na.form`	character representation of `NA`s.
`width`	the returned vector has elements of at most `width`.
`...`	further arguments passed to or from other methods.

Details

These functions convert their first argument to a vector (or array) of character strings which have a common format (as is done by `print`), fulfilling `length(format*(x, *)) == length(x)`. The trimming with `trim = TRUE` is useful when the strings are to be used for plot `axis` annotation.

`format.AsIs` deals with columns of complicated objects that have been extracted from a data frame.

`format.pval` is mainly an auxiliary function for `print.summary.lm` etc., and does separate formatting for fixed, floating point and very small values; those less than `eps` are formatted as `"< [eps]"` (where "[eps]" stands for `format(eps, digits)`.

The function `formatC` provides a rather more flexible formatting facility for numbers, but does *not* provide a common format for several numbers, nor it is platform-independent.

`format.data.frame` formats the data frame column by column, applying the appropriate method of `format` for each column.

`prettyNum` is the utility function for prettifying `x`. If `x` is not a character, `format(x[i], ...)` is applied to each element, and then it is left unchanged if all the other arguments are at their defaults. Note that `prettyNum(x)` may behave unexpectedly if `x` is a `character` not resulting from something like `format(<number>)`.

Note

Currently `format` drops trailing zeroes, so `format(6.001, digits=2)` gives `"6"` and `format(c(6.0, 13.1), digits=2)` gives `c(" 6", "13")`.

Character(s) `"` in input strings `x` are escaped to `\"`.

References

Becker, R. A., Chambers, J. M. and Wilks, A. R. (1988) *The New S Language.* Wadsworth & Brooks/Cole.

See Also

`format.info` indicates how something would be formatted; `formatC`, `paste`, `as.character`, `sprintf`.

Examples

```
format(1:10)

zz <- data.frame("(row names)"= c("aaaaa", "b"),
                 check.names=FALSE)
format(zz)
format(zz, justify="left")

## use of nsmall
format(13.7)
format(13.7, nsmall=3)
```

```
r <- c("76491283764.97430", "29.12345678901", "-7.1234",
       "-100.1","1123")
## American:
prettyNum(r, big.mark = ",")
## Some Europeans:
prettyNum(r, big.mark = "'", decimal.mark = ",")

(dd <- sapply(1:10,
              function(i)paste((9:0)[1:i],collapse="")))
prettyNum(dd, big.mark="'")

pN <- pnorm(1:7, lower=FALSE)
cbind(format (pN, small.mark=" ", digits=15))
cbind(formatC(pN, small.mark=" ", digits=17, format="f"))
```

format.info *format(.) Information*

Description

Information is returned on how `format(x, digits = options("digits"))` would be formatted.

Usage

```
format.info(x, nsmall = 0)
```

Arguments

`x`	(numeric) vector; potential argument of `format(x,...)`.
`nsmall`	(see `format(*, nsmall)`).

Value

An `integer vector` of length 3, say `r`.

`r[1]`	width (number of characters) used for `format(x)`
`r[2]`	number of digits after decimal point.
`r[3]`	in `0:2`; if $\geq$1, *exponential* representation would be used, with exponent length of `r[3]+1`.

Note

The result **depends** on the value of `options("digits")`.

See Also

`format`, `formatC`.

Examples

```
dd <- options("digits") ; options(digits = 7)
# for the following
format.info(123) # 3 0 0
format.info(pi)  # 8 6 0
format.info(1e8) # 5 0 1 - exponential "1e+08"
format.info(1e222) # 6 0 2 - exponential "1e+222"
```

```
x <- pi*10^c(-10,-2,0:2,8,20)
names(x) <- formatC(x,w=1,dig=3,format="g")
cbind(sapply(x,format))
t(sapply(x, format.info))

## using at least 8 digits right of "."
t(sapply(x, format.info, nsmall = 8))

# Reset old options:
options(dd)
```

`formatC` *Formatting Using C-style Formats*

Description

Formatting numbers individually and flexibly, using `C` style format specifications. `format.char` is a helper function for `formatC`.

Usage

```
formatC(x, digits = NULL, width = NULL,
        format = NULL, flag = "", mode = NULL,
        big.mark = "", big.interval = 3,
      small.mark = "", small.interval = 5,
    decimal.mark = ".")

format.char(x, width = NULL, flag = "-")
```

Arguments

`x`	an atomic numerical or character object, typically a vector of real numbers.
`digits`	the desired number of digits after the decimal point (`format = "f"`) or *significant* digits (`format = "g"`, `= "e"` or `= "fg"`). Default: 2 for integer, 4 for real numbers. If less than 0, the C default of 6 digits is used.
`width`	the total field width; if both `digits` and `width` are unspecified, `width` defaults to 1, otherwise to `digits + 1`. `width = 0` will use `width = digits`, `width < 0` means left justify the number in this field (equivalent to `flag ="-"`). If necessary, the result will have more characters than `width`.
`format`	equal to `"d"` (for integers), `"f"`, `"e"`, `"E"`, `"g"`, `"G"`, `"fg"` (for reals), or `"s"` (for strings). Default is `"d"` for integers, `"g"` for reals. `"f"` gives numbers in the usual `xxx.xxx` format; `"e"` and `"E"` give `n.ddde+nn` or `n.dddE+nn` (scientific format); `"g"` and `"G"` put `x[i]` into scientific format only if it saves space to do so. `"fg"` uses fixed format as `"f"`, but `digits` as the minimum number of *significant* digits. That this can lead

to quite long result strings, see examples below. Note that unlike `signif` this prints large numbers with more significant digits than `digits`.

`flag` format modifier as in Kernighan and Ritchie (1988, page 243). `"0"` pads leading zeros; `"-"` does left adjustment, others are `"+"`, `" "`, and `"#"`.

`mode` `"double"` (or `"real"`), `"integer"` or `"character"`. Default: Determined from the storage mode of `x`.

`big.mark, big.interval`
used for prettying longer decimal sequences, passed to `prettyNum`: that help page explains the details.

`small.mark, small.interval, decimal.mark`
as above

Details

If you set `format` it over-rides the setting of `mode`, so `formatC(123.45, mode="double", format="d")` gives `123`.

The rendering of scientific format is platform-dependent: some systems use `n.ddde+nnn` or `n.dddenn` rather than `n.ddde+nn`.

`formatC` does not necessarily align the numbers on the decimal point, so `formatC(c(6.11, 13.1), digits=2, format="fg")` gives `c("6.1", " 13")`. If you want common formatting for several numbers, use `format`.

Value

A character object of same size and attributes as `x`. Unlike `format`, each number is formatted individually. Looping over each element of `x`, `sprintf(...)` is called (inside the C function `str_signif`).

`format.char(x)` and `formatC`, for character `x`, do simple (left or right) padding with white space.

Author(s)

Originally written by Bill Dunlap, later much improved by Martin Maechler, it was first adapted for R by Friedrich Leisch.

References

Kernighan, B. W. and Ritchie, D. M. (1988) *The C Programming Language.* Second edition. Prentice Hall.

See Also

`format`, `sprintf` for more general C like formatting.

Examples

```
xx  <- pi * 10^(-5:4)
cbind(format(xx, digits=4), formatC(xx))
cbind(formatC(xx, wid=9, flag="-"))
cbind(formatC(xx, dig=5, wid=8, format="f", flag="0"))
cbind(format(xx, digits=4), formatC(xx, dig=4, format="fg"))

format.char(c("a", "Abc", "no way"), wid=-7) # <=> flag="-"
formatC(    c("a", "Abc", "no way"), wid=-7) # <=> flag="-"
formatC(c((-1:1)/0,c(1,100)*pi), wid=8, dig=1)

xx <- c(1e-12,-3.98765e-10,1.45645e-69,1e-70,pi*1e37,3.44e4)
##        1          2              3           4         5        6
formatC(xx)
formatC(xx, format="fg")       # special "fixed" format.
formatC(xx, format="f", dig=80) # also long strings
```

formatDL *Format Description Lists*

Description

Format vectors of items and their descriptions as 2-column tables or LaTeX-style description lists.

Usage

```
formatDL(x, y, style = c("table", "list"),
         width = 0.9 * getOption("width"), indent = NULL)
```

Arguments

`x`	a vector giving the items to be described, or a list of length 2 or a matrix with 2 columns giving both items and descriptions.
`y`	a vector of the same length as `x` with the corresponding descriptions. Only used if `x` does not already give the descriptions.
`style`	a character string specifying the rendering style of the description information. If `"table"`, a two-column table with items and descriptions as columns is produced (similar to Texinfo's `@table` environment. If `"list"`, a LaTeX-style tagged description list is obtained.
`width`	a positive integer giving the target column for wrapping lines in the output.
`indent`	a positive integer specifying the indentation of the second column in table style, and the indentation of continuation lines in list style. Must not be greater than `width/2`, and defaults to `width/3` for table style and `width/9` for list style.

Details

After extracting the vectors of items and corresponding descriptions from the arguments, both are coerced to character vectors.

In table style, items with more than `indent - 3` characters are displayed on a line of their own.

Value

a character vector with the formatted entries.

Examples

```
## Use R to create the 'INDEX' for package 'eda' from its
## 'CONTENTS'
x <- read.dcf(file = system.file("CONTENTS",
                                  package = "eda"),
              fields = c("Entry", "Description"))
x <- as.data.frame(x)
writeLines(formatDL(x$Entry, x$Description))
## or equivalently: writeLines(formatDL(x)) Same
## information in tagged description list style:
writeLines(formatDL(x$Entry, x$Description, style = "list"))
## or equivalently: writeLines(formatDL(x, style = "list"))
```

ftable *Flat Contingency Tables*

Description

Create "flat" contingency tables.

Usage

```
ftable(x, ...)

## Default S3 method:
ftable(..., exclude = c(NA, NaN), row.vars = NULL,
       col.vars = NULL)
```

Arguments

`x, ...`	R objects which can be interpreted as factors (including character strings), or a list (or data frame) whose components can be so interpreted, or a contingency table object of class `"table"` or `"ftable"`.
`exclude`	values to use in the exclude argument of `factor` when interpreting non-factor objects.
`row.vars`	a vector of integers giving the numbers of the variables, or a character vector giving the names of the variables to be used for the rows of the flat contingency table.
`col.vars`	a vector of integers giving the numbers of the variables, or a character vector giving the names of the variables to be used for the columns of the flat contingency table.

Details

`ftable` creates "flat" contingency tables. Similar to the usual contingency tables, these contain the counts of each combination of the levels of the variables (factors) involved. This information is then re-arranged as a matrix whose rows and columns correspond to unique combinations of the levels of the row and column variables (as specified by `row.vars` and `col.vars`, respectively). The combinations are created by looping over the variables in reverse order (so that the levels of the "left-most" variable vary the slowest). Displaying a contingency table in this flat matrix form (via `print.ftable`, the print method for objects of class `"ftable"`) is often preferable to showing it as a higher-dimensional array.

`ftable` is a generic function. Its default method, `ftable.default`, first creates a contingency table in array form from all arguments except `row.vars` and `col.vars`. If the first argument is of class `"table"`, it represents a contingency table and is used as is; if it is a flat table of class `"ftable"`, the information it contains is converted to the usual array representation using `as.ftable`. Otherwise, the arguments should be R objects which can be interpreted as factors (including character strings), or a list (or data frame) whose components can be so interpreted, which are cross-tabulated using `table`. Then, the arguments `row.vars` and `col.vars` are used to collapse the contingency table into flat form. If neither of these two is given, the last variable is used for the columns. If both are given and their union is a proper subset of all variables involved, the other variables are summed out.

Function `ftable.formula` provides a formula method for creating flat contingency tables.

Value

`ftable` returns an object of class `"ftable"`, which is a matrix with counts of each combination of the levels of variables with information on the names and levels of the (row and columns) variables stored as attributes `"row.vars"` and `"col.vars"`.

See Also

`ftable.formula` for the formula interface (which allows a `data = .` argument); `read.ftable` for information on reading, writing and coercing flat contingency tables; `table` for "ordinary" cross-tabulation; `xtabs` for formula-based cross-tabulation.

Examples

```
## Start with a contingency table.
data(Titanic)
ftable(Titanic, row.vars = 1:3)
ftable(Titanic, row.vars = 1:2, col.vars = "Survived")
ftable(Titanic, row.vars = 2:1, col.vars = "Survived")

## Start with a data frame.
data(mtcars)
```

```
x <- ftable(mtcars[c("cyl", "vs", "am", "gear")])
x
ftable(x, row.vars = c(2, 4))
```

---

ftable.formula *Formula Notation for Flat Contingency Tables*

Description

Produce or manipulate a flat contingency table using formula notation.

Usage

```
## S3 method for class 'formula':
ftable(formula, data = NULL, subset, na.action, ...)
```

Arguments

`formula`	a formula object with both left and right hand sides specifying the column and row variables of the flat table.
`data`	a data frame, list or environment containing the variables to be cross-tabulated, or a contingency table (see below).
`subset`	an optional vector specifying a subset of observations to be used. Ignored if `data` is a contingency table.
`na.action`	a function which indicates what should happen when the data contain `NA`s. Ignored if `data` is a contingency table.
`...`	further arguments to the default ftable method may also be passed as arguments, see `ftable.default`.

Details

This is a method of the generic function `ftable`.

The left and right hand side of `formula` specify the column and row variables, respectively, of the flat contingency table to be created. Only the `+` operator is allowed for combining the variables. A `.` may be used once in the formula to indicate inclusion of all the "remaining" variables.

If `data` is an object of class `"table"` or an array with more than 2 dimensions, it is taken as a contingency table, and hence all entries should be nonnegative. Otherwise, if it is not a flat contingency table (i.e., an object of class `"ftable"`), it should be a data frame or matrix, list or environment containing the variables to be cross-tabulated. In this case, `na.action` is applied to the data to handle missing values, and, after possibly selecting a subset of the data as specified by the `subset` argument, a contingency table is computed from the variables.

The contingency table is then collapsed to a flat table, according to the row and column variables specified by `formula`.

Value

A flat contingency table which contains the counts of each combination of the levels of the variables, collapsed into a matrix for suitably displaying the counts.

See Also

`ftable`, `ftable.default`; `table`.

Examples

```
data(Titanic)
Titanic
x <- ftable(Survived ~ ., data = Titanic)
x
ftable(Sex ~ Class + Age, data = x)
```

function *Function Definition*

Description

These functions provide the base mechanisms for defining new functions in the R language.

Usage

```
function( arglist ) expr
return(value)
```

Arguments

`arglist`	Empty or one or more name or name=expression terms.
`value`	An expression.

Details

In R (unlike S) the names in an argument list cannot be quoted non-standard names.

If `value` is missing, `NULL` is returned. If it is a single expression, the value of the evaluated expression is returned.

If the end of a function is reached without calling `return`, the value of the last evaluated expression is returned.

Warning

Prior to R 1.8.0, `value` could be a series of non-empty expressions separated by commas. In that case the value returned is a list of the evaluated expressions, with names set to the expressions where these are the names of R objects. That is, `a=foo()` names the list component `a` and gives it value the result of evaluating `foo()`.

This has been deprecated (and a warning is given), as it was never documented in S, and whether or not the list is named differs by S versions.

References

Becker, R. A., Chambers, J. M. and Wilks, A. R. (1988) *The New S Language.* Wadsworth & Brooks/Cole.

See Also

`args` and `body` for accessing the arguments and body of a function.

`debug` for debugging; `invisible` for `return(.)`ing *invisibly*.

Examples

```
norm <- function(x) sqrt(x%*%x)
norm(1:4)

## An anonymous function:
(function(x,y){ z <- x^2 + y^2; x+y+z })(0:7, 1)
```

gc *Garbage Collection*

Description

A call of gc causes a garbage collection to take place. `gcinfo` sets a flag so that automatic collection is either silent (`verbose=FALSE`) or prints memory usage statistics (`verbose=TRUE`).

Usage

```
gc(verbose = getOption("verbose"))
gcinfo(verbose)
```

Arguments

`verbose`	logical; if `TRUE`, the garbage collection prints statistics about cons cells and the vector heap.

Details

A call of `gc` causes a garbage collection to take place. This takes place automatically without user intervention, and the primary purpose of calling `gc` is for the report on memory usage.

However, it can be useful to call `gc` after a large object has been removed, as this may prompt R to return memory to the operating system.

Value

`gc` returns a matrix with rows `"Ncells"` (*cons cells*, usually 28 bytes each on 32-bit systems and 56 bytes on 64-bit systems, and `"Vcells"` (*vector cells*, 8 bytes each), and columns `"used"` and `"gc trigger"`, each also interpreted in megabytes (rounded up to the next 0.1Mb).

If maxima have been set for either `"Ncells"` or `"Vcells"`, a fifth column is printed giving the current limits in Mb (with `NA` denoting no limit).

`gcinfo` returns the previous value of the flag.

See Also

`Memory` on R's memory management and `gctorture` if you are an R hacker.

Examples

```
gc() # do it now
gcinfo(TRUE) # in the future, show when R does it
x <- integer(100000); for(i in 1:18) x <- c(x,i)
gcinfo(verbose = FALSE) # don't show it anymore

gc(TRUE)
```

gc.time *Report Time Spent in Garbage Collection*

Description

This function reports the time spent in garbage collection so far in the R session while GC timing was enabled..

Usage

```
gc.time(on = TRUE)
```

Arguments

on	logical; if TRUE, GC timing is enabled.

Value

A numerical vector of length 5 giving the user CPU time, the system CPU time, the elapsed time and children's user and system CPU times (normally both zero).

Warnings

This is experimental functionality, likely to be removed as soon as the next release.

The timings are rounded up by the sampling interval for timing processes, and so are likely to be over-estimates.

See Also

`gc`, `proc.time` for the timings for the session.

Examples

```
gc.time()
```

`gctorture` *Torture Garbage Collector*

Description

Provokes garbage collection on (nearly) every memory allocation. Intended to ferret out memory protection bugs. Also makes R run *very* slowly, unfortunately.

Usage

```
gctorture(on = TRUE)
```

Arguments

`on`	logical; turning it on/off.

Value

Previous value.

Author(s)

Peter Dalgaard

`get` *Return a Variable's Value*

Description

Search for an R object with a given name and return it if found.

Usage

```
get(x, pos=-1, envir=as.environment(pos), mode="any",
    inherits=TRUE)
```

Arguments

`x`	a variable name (given as a character string).
`pos`	where to look for the object (see the details section); if omitted, the function will search, as if the name of the object appeared in unquoted in an expression.
`envir`	an alternative way to specify an environment to look in; see the details section.
`mode`	the mode of object sought.
`inherits`	should the enclosing frames of the environment be inspected?

Details

The `pos` argument can specify the environment in which to look for the object in any of several ways: as an integer (the position in the `search` list); as the character string name of an element in the search list; or as an `environment` (including using `sys.frame` to access the currently active function calls). The `envir` argument is an alternative way to specify an environment, but is primarily there for back compatibility.

The `mode` includes collections such as `"numeric"` and `"function"`: any member of the collection will suffice.

Value

This function searches the specified environment for a bound variable whose name is given by the character string `x`. If the variable's value is not of the correct `mode`, it is ignored.

If `inherits` is `FALSE`, only the first frame of the specified environment is inspected. If `inherits` is `TRUE`, the search is continued up through the parent frames until a bound value of the right mode is found.

Using a `NULL` environment is equivalent to using the current environment.

References

Becker, R. A., Chambers, J. M. and Wilks, A. R. (1988) *The New S Language.* Wadsworth & Brooks/Cole.

See Also

`exists`.

Examples

```
get("%o%")
```

`getAnywhere` *Retrieve an R Object, Including from a Namespace*

Description

This functions locates all objects with name matching its argument, whether visible on the search path, registered as an S3 method or in a namespace but not exported.

Usage

```
getAnywhere(x)
```

Arguments

`x`	a character string or name.

Details

The function looks at all loaded namespaces, whether or not they are associated with a package on the search list.

Where functions are found as an S3 method, an attempt is made to find which namespace registered them. This may not be correct, especially if a namespace is unloaded.

Value

An object of class `"getAnywhere"`. This is a list with components

`name`	the name searched for.
`funs`	a list of objects found
`where`	a character vector explaining where the object(s) were found
`visible`	logical: is the object visible
`dups`	logical: is the object identical to one earlier in the list.

Normally the structure will be hidden by the `print` method. There is a `[` method to extract one or more of the objects found.

See Also

`get`, `getFromNamespace`

Examples

```
getAnywhere("format.dist")
getAnywhere("simpleLoess") # not exported from modreg
```

`getFromNamespace` *Utility functions for Developing Namespaces*

Description

Utility functions to access and replace the non-exported functions in a namespace, for use in developing packages with namespaces.

Usage

```
getFromNamespace(x, ns, pos=-1, envir=as.environment(pos))
fixInNamespace(x, ns, pos=-1, envir=as.environment(pos), ...)
```

Arguments

`x`	an object name (given as a character string).
`ns`	a namespace, or character string giving the namespace.
`pos`	where to look for the object: see `get`.
`envir`	an alternative way to specify an environment to look in.
`...`	arguments to pass to the editor: see `edit`.

Details

The namespace can be specified in several ways. Using, for example, `ns="modreg"` is the most direct, but a loaded package with a namespace can be specified via any of the methods used for `get`: `ns` can also be the environment `<namespace:foo>`.

`fixInNamespace` invokes `edit` on the object named `x` and assigns the revised object in place of the original object. For compatibility with `fix`, `x` can be unquoted.

Value

`getFromNamespace` returns the object found (or gives an error).

`fixInNamespace` is invoked for its side effect of changing the object in the namespace.

Note

`fixInNamespace` will alter the copy of the object in the namespace, and also a copy registered as an S3 method. There can be other copies, so the function is not foolproof, but should be helpful for debugging.

See Also

`get`, `fix`, `getS3method`

Examples

```
fixInNamespace("predict.ppr", "modreg")
## alternatively
fixInNamespace("predict.ppr", pos = 5)
```

`getNativeSymbolInfo` *Obtain a description of a native (C/Fortran) symbol*

Description

This finds and returns as comprehensive a description of a dynamically loaded or "exported" built-in native symbol. It returns information about the name of the symbol, the library in which it is located and, if available, the number of arguments it expects and by which interface it should be called (i.e `.Call`, `.C`, `.Fortran`, or `.External`). Additionally, it returns the address of the symbol and this can be passed to other C routines which can invoke. Specifically, this provides a way to explicitly share symbols between different dynamically loaded package libraries. Also, it provides a way to query where symbols were resolved, and aids diagnosing strange behavior associated with dynamic resolution.

Usage

```
getNativeSymbolInfo(name, PACKAGE)
```

Arguments

`name`	the name of the native symbol as used in a call to `is.loaded`, etc.
`PACKAGE`	an optional argument that specifies to which dynamically loaded library we restrict the search for this symbol. If this is `"base"`, we search in the R executable itself.

Details

This uses the same mechanism for resolving symbols as is used in all the native interfaces (`.Call`, etc.). If the symbol has been explicitly registered by the shared library in which it is contained, information about the number of arguments and the interface by which it should be called will be returned. Otherwise, a generic native symbol object is returned.

Value

If the symbol is not found, an error is raised. Otherwise, the value is a list containing the following elements:

`name` the name of the symbol, as given by the `name` argument.

`address` the native memory address of the symbol which can be used to invoke the routine, and also compare with other symbol address. This is an external pointer object and of class `NativeSymbol`.

`package` a list containing 3 elements:

- **name** the short form of the library name which can be used as the value of the `PACKAGE` argument in the different native interface functions.
- **path** the fully qualified name of the shared library file.
- **dynamicLookup** a logical value indicating whether dynamic resolution is used when looking for symbols in this library, or only registered routines can be located.

`numParameters` the number of arguments that should be passed in a call to this routine.

Additionally, the list will have an additional class, being `CRoutine`, `CallRoutine`, `FortranRoutine` or `ExternalRoutine` corresponding to the R interface by which it should be invoked.

Note

One motivation for accessing this reflectance information is to be able to pass native routines to C routines as "function pointers" in C. This allows us to treat native routines and R functions in a similar manner, such as when passing an R function to C code that makes callbacks to that function at different points in its computation (e.g., `nls`). Additionally, we can resolve the symbol just once and avoid resolving it repeatedly or using the internal cache. In the future, one may be able to treat `NativeSymbol` objects as directly callback objects.

Author(s)

Duncan Temple Lang

References

For information about registering native routines, see "In Search of C/C++ & FORTRAN Routines", R News, volume 1, number 3, 2001, p20–23 (`http://CRAN.R-project.org/doc/Rnews/`).

See Also

`is.loaded`, `.C`, `.Fortran`, `.External`, `.Call`, `dyn.load`.

Examples

```
library(ctest) # normally loaded
getNativeSymbolInfo("dansari")

library(mva)   # normally loaded
getNativeSymbolInfo(symbol.For("hcass2"))
```

`getNumCConverters` *Management of .C argument conversion list*

Description

These functions provide facilities to manage the extensible list of converters used to translate R objects to C pointers for use in `.C` calls. The number and a description of each element in the list can be retrieved. One can also query and set the activity status of individual elements, temporarily ignoring them. And one can remove individual elements.

Usage

```
getNumCConverters()
getCConverterDescriptions()
getCConverterStatus()
setCConverterStatus(id, status)
removeCConverter(id)
```

Arguments

`id`	either a number or a string identifying the element of interest in the converter list. A string is matched against the description strings for each element to identify the element. Integers are specified starting at 1 (rather than 0).
`status`	a logical value specifying whether the element is to be considered active (`TRUE`) or not (`FALSE`).

Details

The internal list of converters is potentially used when converting individual arguments in a `.C` call. If an argument has a non-trivial class attribute, we iterate over the list of converters looking for the first that "matches". If we find a matching converter, we have it create the C-level pointer corresponding to the R object. When the call to the C routine is complete, we use the same converter for that argument to reverse the conversion and create an R object from the current value in the C pointer. This is done separately for all the arguments.

The functions documented here provide R user-level capabilities for investigating and managing the list of converters. There is currently no mechanism for adding an element to the converter list within the R language. This must be done in C code using the routine `R_addToCConverter()`.

Value

`getNumCConverters` returns an integer giving the number of elements in the list, both active and inactive.

`getCConverterDescriptions` returns a character vector containing the description string of each element of the converter list.

`getCConverterStatus` returns a logical vector with a value for each element in the converter list. Each value indicates whether that converter is active (`TRUE`) or inactive (`FALSE`). The names of the elements are the description strings returned by `getCConverterDescriptions`.

`setCConverterStatus` returns the logical value indicating the activity status of the specified element before the call to change it took effect. This is `TRUE` for active and `FALSE` for inactive.

`removeCConverter` returns `TRUE` if an element in the converter list was identified and removed. In the case that no such element was found, an error occurs.

Author(s)

Duncan Temple Lang

References

`http://developer.R-project.org/CObjectConversion.pdf`

See Also

`.C`

Examples

```
getNumCConverters()
getCConverterDescriptions()
getCConverterStatus()

old <- setCConverterStatus(1,FALSE)

setCConverterStatus(1,old)
```

```
removeCConverter(1)
removeCConverter(getCConverterDescriptions()[1])
```

getpid *Get the Process ID of the R Session*

Description

Get the process ID of the R Session. It is guaranteed by the operating system that two R sessions running simultaneously will have different IDs, but it is possible that R sessions running at different times will have the same ID.

Usage

```
Sys.getpid()
```

Value

An integer, usually a small integer between 0 and 32767 under Unix-alikes and a much small integer under Windows.

Examples

```
Sys.getpid()
```

getS3method *Get An S3 Method*

Description

Get a method for an S3 generic, possibly from a namespace.

Usage

```
getS3method(f, class, optional = FALSE)
```

Arguments

`f`	character: name of the generic.
`class`	character: name of the class.
`optional`	logical: should failure to find the generic or a method be allowed?

Details

S3 methods may be hidden in packages with namespaces, and will not then be found by `get`: this function can retrieve such functions, primarily for debugging purposes.

Value

The function found, or `NULL` if no function is found and `optional = TRUE`.

See Also

`methods`, `get`

Examples

```
require(modreg)
exists("predict.ppr") # false
getS3method("predict", "ppr")
```

`getwd` *Get or Set Working Directory*

Description

`getwd` returns an absolute filename representing the current working directory of the R process; `setwd(dir)` is used to set the working directory to `dir`.

Usage

```
getwd()
setwd(dir)
```

Arguments

`dir`	A character string.

Value

`getwd` returns a character vector, or `NULL` if the working directory is not available on that platform.

`setwd` returns `NULL` invisibly. It will give an error if it does not succeed.

Note

These functions are not implemented on all platforms.

Examples

```
(WD <- getwd())
if (!is.null(WD)) setwd(WD)
```

grep *Pattern Matching and Replacement*

Description

grep searches for matches to pattern (its first argument) within the character vector x (second argument). regexpr does too, but returns more detail in a different format.

sub and gsub perform replacement of matches determined by regular expression matching.

Usage

```
grep(pattern, x, ignore.case = FALSE, extended = TRUE,
     perl = FALSE, value = FALSE, fixed = FALSE)
sub(pattern, replacement, x,
    ignore.case = FALSE, extended = TRUE, perl = FALSE)
gsub(pattern, replacement, x,
     ignore.case = FALSE, extended = TRUE, perl = FALSE)
regexpr(pattern, text,  extended = TRUE, perl = FALSE,
        fixed = FALSE)
```

Arguments

pattern	character string containing a regular expression (or character string for fixed = TRUE) to be matched in the given character vector.
x, text	a character vector where matches are sought.
ignore.case	if FALSE, the pattern matching is *case sensitive* and if TRUE, case is ignored during matching.
extended	if TRUE, extended regular expression matching is used, and if FALSE basic regular expressions are used.
perl	logical. Should perl-compatible regexps be used if available? Has priority over extended.
value	if FALSE, a vector containing the (integer) indices of the matches determined by grep is returned, and if TRUE, a vector containing the matching elements themselves is returned.
fixed	logical. If TRUE, pattern is a string to be matched as is. Overrides all other arguments.
replacement	a replacement for matched pattern in sub and gsub.

Details

The two `*sub` functions differ only in that `sub` replaces only the first occurrence of a `pattern` whereas `gsub` replaces all occurrences.

For `regexpr` it is an error for `pattern` to be `NA`, otherwise `NA` is permitted and matches only itself.

The regular expressions used are those specified by POSIX 1003.2, either extended or basic, depending on the value of the `extended` argument, unless `perl = TRUE` when they are those of PCRE, `ftp://ftp.csx.cam.ac.uk/pub/software/programming/pcre/`. (The exact set of patterns supported may depend on the version of PCRE installed on the system in use.)

Value

For `grep` a vector giving either the indices of the elements of `x` that yielded a match or, if `value` is `TRUE`, the matched elements.

For `sub` and `gsub` a character vector of the same length as the original.

For `regexpr` an integer vector of the same length as `text` giving the starting position of the first match, or −1 if there is none, with attribute `"match.length"` giving the length of the matched text (or −1 for no match).

References

Becker, R. A., Chambers, J. M. and Wilks, A. R. (1988) *The New S Language.* Wadsworth & Brooks/Cole (grep)

See Also

regular expression for the details of the pattern specification.

`agrep` for approximate matching.

`tolower`, `toupper` and `chartr` for character translations. `charmatch`, `pmatch`, `match`. `apropos` uses regexps and has nice examples.

Examples

```
grep("[a-z]", letters)

txt <- c("arm","foot","lefroo", "bafoobar")
if(any(i <- grep("foo",txt)))
   cat("'foo' appears at least once in\n\t",txt,"\n")
i # 2 and 4
```

```
txt[i]

## Double all 'a' or 'b's;  "\" must be escaped, i.e.,
## 'doubled'
gsub("([ab])", "\\1_\\1_", "abc and ABC")

txt <- c("The", "licenses", "for", "most", "software",
  "are", "designed", "to", "take", "away", "your",
  "freedom", "to", "share", "and", "change", "it.",
   "", "By", "contrast,", "the", "GNU", "General", "Public",
   "License", "is", "intended", "to", "guarantee", "your",
   "freedom", "to", "share", "and", "change", "free",
   "software", "--", "to", "make", "sure", "the",
   "software", "is", "free", "for", "all", "its", "users")
( i <- grep("[gu]", txt) ) # indices
stopifnot( txt[i] == grep("[gu]", txt, value = TRUE) )
(ot <- sub("[b-e]",".", txt))
# gsub does "global" substitution
txt[ot != gsub("[b-e]",".", txt)]

txt[gsub("g","#", txt) !=
    gsub("g","#", txt, ignore.case = TRUE)] # the "G" words

regexpr("en", txt)

## trim trailing white space
str = 'Now is the time      '
sub(' +$', '', str)  ## spaces only
sub('[[:space:]]+$', '', str) ## white space, POSIX-style
if(capabilities("PCRE"))
  sub('\\s+$', '', str, perl = TRUE) ## perl-style white
                                     ## space
```

`groupGeneric` *Group Generic Functions*

Description

Group generic functions can be defined with either S3 and S4 methods (with different groups). Methods are defined for the group of functions as a whole.

A method defined for an individual member of the group takes precedence over a method defined for the group as a whole.

When package **methods** is attached there are objects visible with the names of the group generics: these functions should never be called directly (a suitable error message will result if they are).

Usage

```
## S3 methods have prototypes:
Math(x, ...)
Ops(e1, e2)
Summary(x, ...)
Complex(z)

## S4 methods have prototypes:
Arith(e1, e2)
Compare(e1, e2)
Ops(e1, e2)
Math(x)
Math2(x, digits)
Summary(x, ..., na.rm = FALSE)
Complex(z)
```

Arguments

`x, z, e1, e2`	objects.
`digits`	number of digits to be used in `round` or `signif`.
`...`	further arguments passed to or from methods.
`na.rm`	logical: should missing values be removed?

S3 Group Dispatching

There are four *groups* for which S3 methods can be written, namely the `"Math"`, `"Ops"`, `"Summary"` and `"Complex"` groups. These are not R objects, but methods can be supplied for them and base R contains `factor` and `data.frame` methods for the first three groups. (There is also a `ordered` method for `Ops`.)

1. Group `"Math"`:
   - `abs`, `sign`, `sqrt`,
     `floor`, `ceiling`, `trunc`,
     `round`, `signif`
   - `exp`, `log`,
     `cos`, `sin`, `tan`,
     `acos`, `asin`, `atan`
     `cosh`, `sinh`, `tanh`,
     `acosh`, `asinh`, `atanh`
   - `lgamma`, `gamma`, `gammaCody`,
     `digamma`, `trigamma`, `tetragamma`, `pentagamma`
   - `cumsum`, `cumprod`, `cummax`, `cummin`
2. Group `"Ops"`:
   - `"+"`, `"-"`, `"*"`, `"/"`, `"^"`, `"%%"`, `"%/%"`
   - `"&"`, `"|"`, `"!"`
   - `"=="`, `"!="`, `"<"`, `"<="`, `">="`, `">"`
3. Group `"Summary"`:
   - `all`, `any`
   - `sum`, `prod`
   - `min`, `max`
   - `range`
4. Group `Complex`:
   - `Arg`, `Conj`, `Im`, `Mod`, `Re`

The number of arguments supplied for `"Math"` group generic methods is not checked prior to dispatch. (Prior to R 1.7.0, all those whose default method has one argument were checked, but the others were not.)

S4 Group Dispatching

When package **methods** is attached, formal (S4) methods can be defined for groups.

The functions belonging to the various groups are as follows:

Arith "+", "-", "*", "^", "%%", "%/%", "/"

Compare "==", ">", "<", "!=", "<=", ">="

Ops "Arith", "Compare"

Math "log", "sqrt", "log10", "cumprod", "abs", "acos", "acosh", "asin", "asinh", "atan", "atanh", "ceiling", "cos", "cosh", "cumsum", "exp", "floor", "gamma", "lgamma", "sin", "sinh", "tan", "tanh", "trunc"

Math2 "round", "signif"

Summary "max", "min", "range", "prod", "sum", "any", "all"

Complex "Arg", "Conj", "Im", "Mod", "Re"

Functions with the group names exist in the **methods** package but should not be called directly.

All the functions in these groups (other than the group generics themselves) are basic functions in R. They are not by default S4 generic functions, and many of them are defined as primitives, meaning that they do not have formal arguments. However, you can still define formal methods for them. The effect of doing so is to create an S4 generic function with the appropriate arguments, in the environment where the method definition is to be stored. It all works more or less as you might expect, admittedly via a bit of trickery in the background.

References

Appendix A, *Classes and Methods* of
Chambers, J. M. and Hastie, T. J. eds (1992) *Statistical Models in S.* Wadsworth & Brooks/Cole.

Chambers, J. M. (1998) *Programming with Data.* Springer, pp. 352–4.

See Also

`methods` for methods of non-Internal generic functions.

Examples

```
methods("Math")
methods("Ops")
methods("Summary")

d.fr <- data.frame(x=1:9, y=rnorm(9))
data.class(1 + d.fr) == "data.frame" ## add to d.f. ...
```

gzcon *(De)compress I/O Through Connections*

Description

gzcon provides a modified connection that wraps an existing connection, and decompresses reads or compresses writes through that connection. Standard gzip headers are assumed.

Usage

```
gzcon(con, level = 6, allowNonCompressed = TRUE)
```

Arguments

`con` a connection.

`level` integer between 0 and 9, the compression level when writing.

`allowNonCompressed` logical. When reading, should non-compressed files (lacking the gzip magic header) be allowed?

Details

If con is open then the modified connection is opened. Closing the wrapper connection will also close the underlying connection.

Reading from a connection which does not supply a gzip magic header is equivalent to reading from the original connection if allowNonCompressed is true, otherwise an error.

The original connection is unusable: any object pointing to it will now refer to the modified connection.

Value

An object inheriting from class "connection". This is the same connection *number* as supplied, but with a modified internal structure.

See Also

gzfile

Examples

```
## print the value to see what objects were created.
con <- url("http://host/file.sav")
print(load(con))

## gzfile and gzcon can inter-work. Of course here one
## would used gzfile, but file() can be replaced by any
## other connection generator.
zz <- gzfile("ex.gz", "w")
cat("TITLE extra line", "2 3 5 7", "", "11 13 17",
    file = zz, sep = "\n")
close(zz)
readLines(zz<-gzcon(file("ex.gz")))
close(zz)
unlink("ex.gz")

zz <- gzcon(file("ex.gz", "wb"))
cat("TITLE extra line", "2 3 5 7", "", "11 13 17",
    file = zz, sep = "\n")
close(zz)
readLines(zz<-gzfile("ex.gz"))
close(zz)
unlink("ex.gz")
```

help *Documentation*

Description

These functions provide access to documentation. Documentation on a topic with name `name` (typically, an R object or a data set) can be printed with either `help(name)` or `?name`.

Usage

```
help(topic, offline = FALSE, package = .packages(),
     lib.loc = NULL, verbose = getOption("verbose"),
     try.all.packages = getOption("help.try.all.packages"),
     htmlhelp = getOption("htmlhelp"),
     pager = getOption("pager"))
?topic
type?topic
```

Arguments

`topic` usually, the name on which documentation is sought. The name may be quoted or unquoted (but note that if `topic` is the name of a variable containing a character string documentation is provided for the name, not for the character string).

The `topic` argument may also be a function call, to ask for documentation on a corresponding method. See the section on method documentation.

`offline` a logical indicating whether documentation should be displayed on-line to the screen (the default) or hardcopy of it should be produced.

`package` a name or character vector giving the packages to look into for documentation. By default, all packages in the search path are used.

`lib.loc` a character vector of directory names of R libraries, or `NULL`. The default value of `NULL` corresponds to all libraries currently known. If the default is used, the loaded packages are searched before the libraries.

`verbose` logical; if `TRUE`, the file name is reported.

`try.all.packages`
: logical; see `Notes`.

`htmlhelp`
: logical (or `NULL`). If `TRUE` (which is the default after `help.start` has been called), the HTML version of the help will be shown in the browser specified by `options("browser")`. See `browseURL` for details of the browsers that are supported. Where possible an existing browser window is re-used.

`pager`
: the pager to be used for `file.show`.

`type`
: the special type of documentation to use for this topic; for example, if the type is `class`, documentation is provided for the class with name `topic`. The function `topicName` returns the actual name used in this case. See the section on method documentation for the uses of `type` to get help on formal methods.

Details

In the case of unary and binary operators and control-flow special forms (including `if`, `for` and `function`), the topic may need to be quoted.

If `offline` is `TRUE`, hardcopy of the documentation is produced by running the LaTeX version of the help page through `latex` (note that LaTeX 2e is needed) and `dvips`. Depending on your `dvips` configuration, hardcopy will be sent to the printer or saved in a file. If the programs are in non-standard locations and hence were not found at compile time, you can either set the options `latexcmd` and `dvipscmd`, or the environment variables `R_LATEXCMD` and `R_DVIPSCMD` appropriately. The appearance of the output can be customized through a file '`Rhelp.cfg`' somewhere in your LaTeX search path.

Method Documentation.

The authors of formal ('S4') methods can provide documentation on specific methods, as well as overall documentation on the methods of a particular function. The `"?"` operator allows access to this documentation in three ways.

The expression `methods ? f` will look for the overall documentation methods for the function `f`. Currently, this means the documentation file containing the alias `f-methods`.

There are two different ways to look for documentation on a particular method. The first is to supply the `topic` argument in the form of a function call, omitting the `type` argument. The effect is to look for documentation on the method that would be used if this function call were actually evaluated. See the examples below. If the function is not a generic (no S4 methods are defined for it), the help reverts to documentation on the function name.

The `"?"` operator can also be called with `type` supplied as `"method"`; in this case also, the `topic` argument is a function call, but the arguments are now interpreted as specifying the class of the argument, not the actual expression that will appear in a real call to the function. See the examples below.

The first approach will be tedious if the actual call involves complicated expressions, and may be slow if the arguments take a long time to evaluate. The second approach avoids these difficulties, but you do have to know what the classes of the actual arguments will be when they are evaluated.

Both approaches make use of any inherited methods; the signature of the method to be looked up is found by using `selectMethod` (see the documentation for `getMethod`).

Note

Unless `lib.loc` is specified explicitly, the loaded packages are searched before those in the specified libraries. This ensures that if a library is loaded from a library not in the known library trees, then the help from the loaded library is used. If `lib.loc` is specified explicitly, the loaded packages are *not* searched.

If this search fails and argument `try.all.packages` is `TRUE` and neither `packages` nor `lib.loc` is specified, then all the packages in the known library trees are searched for help on `topic` and a list of (any) packages where help may be found is printed (but no help is shown). **N.B.** searching all packages can be slow.

The help files can be many small files. On some file systems it is desirable to save space, and the text files in the '`help`' directory of an installed package can be zipped up as a zip archive '`Rhelp.zip`'. Ensure that file '`AnIndex`' remains un-zipped. Similarly, all the files in the '`latex`' directory can be zipped to '`Rhelp.zip`'.

References

Becker, R. A., Chambers, J. M. and Wilks, A. R. (1988) *The New S Language*. Wadsworth & Brooks/Cole.

See Also

`help.search()` for finding help pages on a "vague" topic; `help.start()` which opens the HTML version of the R help pages; `library()` for listing available packages and the user-level objects they contain; `data()` for listing available data sets; `methods()`.

See `prompt()` to get a prototype for writing `help` pages of private packages.

Examples

```
help()
help(help)      # the same

help(lapply)
?lapply         # the same

help("for")     # or ?"for", but the quotes are needed
?"+"

# get help even when package is not loaded
help(package="stepfun")

data()          # list all available data sets
?women          # information about data set "women"

topi <- "women"
help(topi) # Error: No documentation for 'topi'

# reports not found (an error)
try(help("bs", try.all.packages=FALSE))
# reports can be found in package 'splines'
help("bs", try.all.packages=TRUE)

## define a generic function and some methods
combo <- function(x, y) c(x, y)
setGeneric("combo")
setMethod("combo", c("numeric", "numeric"),
   function(x, y) x+y)

## assume we have written some documentation for combo, and
## its methods ....
```

```
## produces the function documentation
?combo
## looks for the overall methods documentation
methods?combo
## documentation for the method above
method?combo("numeric", "numeric")
## ... the same method, selected according to
## the arguments (one integer, the other numeric)
?combo(1:10, rnorm(10))
## documentation for the default method
?combo(1:10, letters)
```

help.search *Search the Help System*

Description

Allows for searching the help system for documentation matching a given character string in the (file) name, alias, title, or keyword entries (or any combination thereof), using either fuzzy matching or regular expression matching. Names and titles of the matched help entries are displayed nicely.

Usage

```
help.search(pattern,
            fields = c("alias", "concept", "title"),
            apropos, keyword, whatis, ignore.case = TRUE,
            package = NULL, lib.loc = NULL,
            help.db = getOption("help.db"),
            verbose = getOption("verbose"),
            rebuild = FALSE, agrep = NULL)
```

Arguments

`pattern`	a character string to be matched in the specified fields. If this is given, the arguments `apropos`, `keyword`, and `whatis` are ignored.
`fields`	a character vector specifying the fields of the help databases to be searched. The entries must be abbreviations of `"name"`, `"title"`, `"alias"`, `"concept"`, and `"keyword"`, corresponding to the help page's (file) name, its title, the topics and concepts it provides documentation for, and the keywords it can be classified to.
`apropos`	a character string to be matched in the help page topics and title.
`keyword`	a character string to be matched in the help page 'keywords'. 'Keywords' are really categories: the standard categories are listed in file '`RHOME/doc/KEYWORDS`' and some package writers have defined their own. If `keyword` is specified, `agrep` defaults to `FALSE`.
`whatis`	a character string to be matched in the help page topics.

`ignore.case` a logical. If `TRUE`, case is ignored during matching; if `FALSE`, pattern matching is case sensitive.

`package` a character vector with the names of packages to search through, or `NULL` in which case *all* available packages in the library trees specified by `lib.loc` are searched.

`lib.loc` a character vector describing the location of R library trees to search through, or `NULL`. The default value of `NULL` corresponds to all libraries currently known.

`help.db` a character string giving the file path to a previously built and saved help database, or `NULL`.

`verbose` logical; if `TRUE`, the search process is traced.

`rebuild` a logical indicating whether the help database should be rebuilt.

`agrep` if `NULL` (the default unless `keyword` is used) and the character string to be matched consists of alphanumeric characters, whitespace or a dash only, approximate (fuzzy) matching via `agrep` is used unless the string has fewer than 5 characters; otherwise, it is taken to contain a regular expression to be matched via `grep`. If `FALSE`, approximate matching is not used. Otherwise, one can give a numeric or a list specifying the maximal distance for the approximate match, see argument `max.distance` in the documentation for `agrep`.

Details

Upon installation of a package, a contents database which contains the information on name, title, aliases and keywords and, concepts starting with R 1.8.0, is computed from the Rd files in the package and serialized as '`Rd.rds`' in the '`Meta`' subdirectory of the top-level package installation directory (or, prior to R 1.7.0, as '`CONTENTS`' in Debian Control Format with aliases and keywords collapsed to character strings in the top-level package installation directory). This, or a pre-built help.search index serialized as '`hsearch.rds`' in the '`Meta`' directory, is the database searched by `help.search()`.

The arguments `apropos` and `whatis` play a role similar to the Unix commands with the same names.

If possible, the help data base is saved to the file '`help.db`' in the '`.R`' subdirectory of the user's home directory or the current working directory.

Note that currently, the aliases in the matching help files are not displayed.

Value

The results are returned in an object of class `"hsearch"`, which has a print method for nicely displaying the results of the query. This mechanism is experimental, and may change in future versions of R.

See Also

`help`; `help.start` for starting the hypertext (currently HTML) version of R's online documentation, which offers a similar search mechanism.

`apropos` uses regexps and has nice examples.

Examples

```
# In case you forgot how to fit linear models
help.search("linear models")

help.search("non-existent topic")

# All help pages with topics or title matching 'print'
help.search("print")
# The same
help.search(apropos = "print")
# All help pages documenting high-level plots.
help.search(keyword = "hplot")

## Help pages with documented topics starting with 'try'.
help.search("\\btry", fields = "alias")
## Do not use '^' or '$' when matching aliases or keywords
## (unless all packages were installed using R 1.7 or
## newer).
```

help.start *Hypertext Documentation*

Description

Start the hypertext (currently HTML) version of R's online documentation.

Usage

```
help.start(gui = "irrelevant",
           browser = getOption("browser"),
           remote = NULL)
```

Arguments

`gui`	just for compatibility with S-PLUS.
`browser`	the name of the program to be used as hypertext browser. It should be in the PATH, or a full path specified.
`remote`	A character giving a valid URL for the '`$R_HOME`' directory on a remote location.

Details

All the packages in the known library trees are linked to directory '`.R`' in the per-session temporary directory. The links are re-made each time `help.start` is run, which should be done after packages are installed, updated or removed.

If the browser given by the `browser` argument is different from the default browser as specified by `options("browser")`, the default is changed to the given browser so that it gets used for all future help requests.

See Also

`help()` for on- and off-line help in ASCII/Editor or PostScript format.

`browseURL` for how the help file is displayed.

Examples

```
help.start()
```

identical *Test Objects for Exact Equality*

Description

The safe and reliable way to test two objects for being *exactly* equal. It returns `TRUE` in this case, `FALSE` in every other case.

Usage

```
identical(x, y)
```

Arguments

`x, y`	any R objects.

Details

A call to `identical` is the way to test exact equality in `if` and `while` statements, as well as in logical expressions that use `&&` or `||`. In all these applications you need to be assured of getting a single logical value.

Users often use the comparison operators, such as `==` or `!=`, in these situations. It looks natural, but it is not what these operators are designed to do in R. They return an object like the arguments. If you expected `x` and `y` to be of length 1, but it happened that one of them wasn't, you will *not* get a single `FALSE`. Similarly, if one of the arguments is `NA`, the result is also `NA`. In either case, the expression `if(x == y)..` `..` won't work as expected.

The function `all.equal` is also sometimes used to test equality this way, but it was intended for something different. First, it tries to allow for "reasonable" differences in numeric results. Second, it returns a descriptive character vector instead of `FALSE` when the objects do not match. Therefore, it is not the right function to use for reliable testing either. (If you *do* want to allow for numeric fuzziness in comparing objects, you can combine `all.equal` and `identical`, as shown in the examples below.)

The computations in `identical` are also reliable and usually fast. There should never be an error. The only known way to kill `identical` is by having an invalid pointer at the C level, generating a memory fault. It will usually find inequality quickly. Checking equality for two large, complicated objects can take longer if the objects are identical or nearly so, but represent completely independent copies. For most applications, however, the computational cost should be negligible.

As from R 1.6.0, `identical` sees `NaN` as different from `as.double(NA)`, but all `NaN`s are equal (and all `NA` of the same type are equal).

Value

A single logical value, `TRUE` or `FALSE`, never `NA` and never anything other than a single value.

Author(s)

John Chambers

References

Chambers, J. M. (1998) *Programming with Data. A Guide to the S Language.* Springer.

See Also

`all.equal` for descriptions of how two objects differ; Comparison for operators that generate elementwise comparisons.

Examples

```
identical(1, NULL) # FALSE -- don't try this with ==
identical(1, 1.)   # TRUE in R (both are stored as doubles)
identical(1, as.integer(1)) # FALSE, stored as different
                            # types

x <- 1.0; y <- 0.99999999999
## how to test for object equality allowing for numeric
## fuzz
identical(all.equal(x, y), TRUE)
## If all.equal thinks the objects are different, it
## returns a character string, and this expression
## evaluates to FALSE

# even for unusual R objects :
identical(.GlobalEnv, environment())
```

`ifelse` *Conditional Element Selection*

Description

`ifelse` returns a value with the same shape as `test` which is filled with elements selected from either `yes` or `no` depending on whether the element of `test` is `TRUE` or `FALSE`. If `yes` or `no` are too short, their elements are recycled.

Usage

```
ifelse(test, yes, no)
```

Arguments

`test`	a logical vector
`yes`	return values for true elements of `test`.
`no`	return values for false elements of `test`.

References

Becker, R. A., Chambers, J. M. and Wilks, A. R. (1988) *The New S Language.* Wadsworth & Brooks/Cole.

See Also

`if`.

Examples

```
x <- c(6:-4)
sqrt(x) # gives warning
sqrt(ifelse(x >= 0, x, NA)) # no warning

## Note: the following also gives the warning !
ifelse(x >= 0, sqrt(x), NA)
```

`index.search` *Search Indices for Help Files*

Description

Used to search the indices for help files, possibly under aliases.

Usage

```
index.search(topic, path, file="AnIndex", type = "help")
```

Arguments

`topic`	The keyword to be searched for in the indices.
`path`	The path(s) to the packages to be searched.
`file`	The index file to be searched. Normally '`"AnIndex"`'.
`type`	The type of file required.

Details

For each package in `path`, examine the file `file` in directory '`type`', and look up the matching file stem for topic `topic`, if any.

Value

A character vector of matching files, as if they are in directory `type` of the corresponding package. In the special cases of `type = "html"`, `"R-ex"` and `"latex"` the file extensions `".html"`, `".R"` and `".tex"` are added.

See Also

`help`, `example`

INSTALL *Install Add-on Packages*

Description

Utility for installing add-on packages.

Usage

```
R CMD INSTALL [options] [-l lib] pkgs
```

Arguments

pkgs	A list with the path names of the packages to be installed.
lib	the path name of the R library tree to install to.
options	a list of options through which in particular the process for building the help files can be controlled.

Details

If used as `R CMD INSTALL pkgs` without explicitly specifying `lib`, packages are installed into the library tree rooted at the first directory given in the environment variable `R_LIBS` if this is set and non-null, and to the default library tree (which is rooted at '`$R_HOME/library`') otherwise.

To install into the library tree `lib`, use `R CMD INSTALL -l lib pkgs`.

Both `lib` and the elements of `pkgs` may be absolute or relative path names. `pkgs` can also contain name of package archive files of the form '`pkg_version.tar.gz`' as obtained from CRAN, these are then extracted in a temporary directory.

Some package sources contain a '`configure`' script that can be passed arguments or variables via the option '`--configure-args`' and '`--configure-vars`', respectively, if necessary. The latter is useful in particular if libraries or header files needed for the package are in non-system directories. In this case, one can use the configure variables `LIBS` and `CPPFLAGS` to specify these locations (and set these via '`--configure-vars`'), see section "Configuration variables" in "R Installation and Administration" for more information. One can also bypass the configure mechanism using the option '`--no-configure`'.

If '`--no-docs`' is given, no help files are built. Options '`--no-text`', '`--no-html`', and '`--no-latex`' suppress creating the text, HTML, and

LaTeX versions, respectively. The default is to build help files in all three versions.

If the option '`--save`' is used, the installation procedure creates a binary image of the package code, which is then loaded when the package is attached, rather than evaluating the package source at that time. Having a file '`install.R`' in the package directory makes this the default behavior for the package (option '`--no-save`' overrides). You may need '`--save`' if your package requires other packages to evaluate its own source. If the file '`install.R`' is non-empty, it should contain R expressions to be executed when the package is attached, after loading the saved image. Options to be passed to R when creating the save image can be specified via '`--save=ARGS`'.

If the attempt to install the package fails, leftovers are removed. If the package was already installed, the old version is restored.

Use `R CMD INSTALL --help` for more usage information.

Packages using the methods package

Packages that require the methods package, and that use functions such as `setMethod` or `setClass`, should be installed by creating a binary image.

The presence of a file named '`install.R`' in the package's main directory causes an image to be saved. Note that the file is not in the '`R`' subdirectory: all the code in that subdirectory is used to construct the binary image.

Normally, the file '`install.R`' will be empty; if it does contain R expressions these will be evaluated when the package is attached, e.g. by a call to the function `library`. (Specifically, the source code evaluated for a package with a saved image consists of a suitable definition of `.First.lib` to ensure loading of the saved image, followed by the R code in file '`install.R`', if any.)

See Also

`REMOVE`, `update.packages` for automatic update of packages using the internet; the chapter on "Creating R packages" in "Writing R Extensions" (see the '`doc/manual`' subdirectory of the R source tree).

`integer` *Integer Vectors*

Description

Creates or tests for objects of type `"integer"`.

Usage

```
integer(length = 0)
as.integer(x, ...)
is.integer(x)
```

Arguments

`length`	desired length.
`x`	object to be coerced or tested.
`...`	further arguments passed to or from other methods.

Value

`integer` creates an integer vector of the specified length. Each element of the vector is equal to `0`. Integer vectors exist so that data can be passed to C or Fortran code which expects them.

`as.integer` attempts to coerce its argument to be of integer type. The answer will be `NA` unless the coercion succeeds. Real values larger in modulus than the largest integer are coerced to `NA` (unlike S which gives the most extreme integer of the same sign). Non-integral numeric values are truncated towards zero (i.e., `as.integer(x)` equals `trunc(x)` there), and imaginary parts of complex numbers are discarded (with a warning). Like `as.vector` it strips attributes including names.

`is.integer` returns `TRUE` or `FALSE` depending on whether its argument is of integer type or not. `is.integer` is generic: you can write methods to handle specific classes of objects, see InternalMethods. Note that factors are true for `is.integer` but false for `is.numeric`.

References

Becker, R. A., Chambers, J. M. and Wilks, A. R. (1988) *The New S Language.* Wadsworth & Brooks/Cole.

See Also

`round` (and `ceiling` and `floor` on that help page) to convert to integral values.

Examples

```
## as.integer() truncates:
x <- pi * c(-1:1,10)
as.integer(x)
```

interaction *Compute Factor Interactions*

Description

`interaction` computes a factor which represents the interaction of the given factors. The result of `interaction` is always unordered.

Usage

```
interaction(..., drop = FALSE)
```

Arguments

`...`	The factors for which interaction is to be computed, or a single list giving those factors.
`drop`	If `drop` is `TRUE`, empty factor levels are dropped from the result. The default is to retain all factor levels.

Value

A factor which represents the interaction of the given factors.

References

Chambers, J. M. and Hastie, T. J. (1992) *Statistical Models in S.* Wadsworth & Brooks/Cole.

See Also

`factor`.

Examples

```
a <- gl(2, 2, 8)
b <- gl(2, 4, 8)
interaction(a, b)
```

interactive *Is R Running Interactively?*

Description

Return TRUE when R is being used interactively and FALSE otherwise.

Usage

```
interactive()
```

See Also

source, .First

Examples

```
.First <- function() if(interactive()) x11()
```

`Internal` *Call an Internal Function*

Description

`.Internal` performs a call to an internal code which is built in to the R interpreter. Only true R wizards should even consider using this function.

Usage

```
.Internal(call)
```

Arguments

`call`	a call expression

See Also

`.Primitive`, `.C`, `.Fortran`.

InternalMethods *Internal Generic Functions*

Description

Many R-internal functions are *generic* and allow methods to be written for.

Details

The following builtin functions are *generic* as well, i.e., you can write methods for them:

[, [[, $, [<-, [[<-, $<-,

length,

dimnames<-, dimnames, dim<-, dim

c, unlist,

as.character, as.vector, is.array, is.atomic, is.call, is.character, is.complex, is.double, is.environment, is.function, is.integer, is.language, is.logical, is.list, is.matrix, is.na, is.nan is.null, is.numeric, is.object, is.pairlist, is.recursive, is.single, is.symbol.

See Also

methods for the methods of non-Internal generic functions.

`invisible` *Change the Print Mode to Invisible*

Description

Return a (temporarily) invisible copy of an object.

Usage

```
invisible(x)
```

Arguments

`x`	an arbitrary R object.

Details

This function can be useful when it is desired to have functions return values which can be assigned, but which do not print when they are not assigned.

References

Becker, R. A., Chambers, J. M. and Wilks, A. R. (1988) *The New S Language.* Wadsworth & Brooks/Cole.

See Also

`return`, `function`.

Examples

```
# These functions both return their argument
f1 <- function(x) x
f2 <- function(x) invisible(x)
f1(1) # prints
f2(1) # does not
```

IQR *The Interquartile Range*

Description

computes interquartile range of the `x` values.

Usage

```
IQR(x, na.rm = FALSE)
```

Arguments

`x`	a numeric vector.
`na.rm`	logical. Should missing values be removed?

Details

Note that this function computes the quartiles using the `quantile` function rather than following Tukey's recommendations, i.e., `IQR(x) = quantile(x,3/4) - quantile(x,1/4)`.

For normally $N(m,1)$ distributed X, the expected value of `IQR(X)` is `2*qnorm(3/4) = 1.3490`, i.e., for a normal-consistent estimate of the standard deviation, use `IQR(x) / 1.349`.

References

Tukey, J. W. (1977). *Exploratory Data Analysis.* Reading: Addison-Wesley.

See Also

`fivenum`, `mad` which is more robust, `range`, `quantile`.

Examples

```
data(rivers)
IQR(rivers)
```

`is.finite` *Finite, Infinite and NaN Numbers*

Description

`is.finite` and `is.infinite` return a vector of the same length as `x`, indicating which elements are finite (not infinite and not missing).

`Inf` and `-Inf` are positive and negative "infinity" whereas `NaN` means "Not a Number".

Usage

```
is.finite(x)
is.infinite(x)
Inf
NaN
is.nan(x)
```

Arguments

`x`	(numerical) object to be tested.

Details

`is.finite` returns a vector of the same length as `x` the jth element of which is `TRUE` if `x[j]` is finite (i.e., it is not one of the values `NA`, `NaN`, `Inf` or `-Inf`). All elements of character and generic (list) vectors are false, so `is.finite` is only useful for logical, integer, numeric and complex vectors. Complex numbers are finite if both the real and imaginary parts are.

`is.infinite` returns a vector of the same length as `x` the jth element of which is `TRUE` if `x[j]` is infinite (i.e., equal to one of `Inf` or `-Inf`).

`is.nan` tests if a numeric value is `NaN`. Do not test equality to `NaN`, or even use `identical`, since systems typically have many different NaN values. In most ports of R one of these is used for the numeric missing value `NA`. It is generic: you can write methods to handle specific classes of objects, see InternalMethods.

Note

In R, basically all mathematical functions (including basic `Arithmetic`), are supposed to work properly with ±`Inf` and `NaN` as input or output.

The basic rule should be that calls and relations with `Inf`s really are statements with a proper mathematical *limit*.

References

ANSI/IEEE 754 Floating-Point Standard.

See Also

`NA`, '*Not Available*' which is not a number as well, however usually used for missing values and applies to many modes, not just numeric.

Examples

```
pi / 0 # = Inf, non-zero number divided by 0 is infinity
0 / 0  # = NaN

1/0 + 1/0 # Inf
1/0 - 1/0 # NaN

stopifnot(
    1/0 == Inf,
    1/Inf == 0
)
sin(Inf)
cos(Inf)
tan(Inf)
```

is.function *Is an Object of Type Function?*

Description

Checks whether its argument is a function.

Usage

```
is.function(x)
```

Arguments

`x`	an R object.

Details

`is.function` is generic: you can write methods to handle specific classes of objects, see InternalMethods.

Value

`TRUE` if `x` is a function, and `FALSE` otherwise.

is.language *Is an Object a Language Object?*

Description

is.language returns TRUE if x is either a variable name, a call, or an expression.

Usage

```
is.language(x)
```

Arguments

x	object to be tested.

Details

is.language is generic: you can write methods to handle specific classes of objects, see InternalMethods.

References

Becker, R. A., Chambers, J. M. and Wilks, A. R. (1988) *The New S Language*. Wadsworth & Brooks/Cole.

Examples

```
ll <- list(a = expression(x^2 - 2*x + 1),
           b = as.name("Jim"),
           c = as.expression(exp(1)),
           d = call("sin", pi))
sapply(ll, typeof)
sapply(ll, mode)
stopifnot(sapply(ll, is.language))
```

is.object *Is an Object "internally classed"?*

Description

A function rather for internal use. It returns `TRUE` if the object `x` has the R internal `OBJECT` attribute set, and `FALSE` otherwise.

Usage

```
is.object(x)
```

Arguments

`x`	object to be tested.

Details

`is.object` is generic: you can write methods to handle specific classes of objects, see InternalMethods.

See Also

`class`, and `methods`.

Examples

```
is.object(1) # FALSE
is.object(as.factor(1:3)) # TRUE
```

is.R *Are we using R, rather than S?*

Description

Test if running under R.

Usage

```
is.R()
```

Details

The function has been written such as to correctly run in all versions of R, S and S-PLUS. In order for code to be runnable in both R and S dialects, either your the code must define `is.R` or use it as

```
if (exists("is.R") && is.function(is.R) && is.R()) {
## R-specific code
} else {
## S-version of code
}
```

Value

`is.R` returns `TRUE` if we are using R and `FALSE` otherwise.

See Also

`R.version`, `system`.

Examples

```
x <- runif(20); small <- x < 0.4
## 'which()' only exists in R:
if(is.R()) which(small) else seq(along=small)[small]
```

is.recursive *Is an Object Atomic or Recursive?*

Description

`is.atomic` returns `TRUE` if `x` does not have a list structure and `FALSE` otherwise.

`is.recursive` returns `TRUE` if `x` has a recursive (list-like) structure and `FALSE` otherwise.

Usage

```
is.atomic(x)
is.recursive(x)
```

Arguments

`x`	object to be tested.

Details

These are generic: you can write methods to handle specific classes of objects, see InternalMethods.

References

Becker, R. A., Chambers, J. M. and Wilks, A. R. (1988) *The New S Language*. Wadsworth & Brooks/Cole.

See Also

`is.list`, `is.language`, etc, and the `demo("is.things")`.

Examples

```
is.a.r <- function(x) c(is.atomic(x), is.recursive(x))

is.a.r(c(a=1,b=3))       # TRUE FALSE
is.a.r(list())           # FALSE TRUE ??
is.a.r(list(2))          # FALSE TRUE
is.a.r(lm)               # FALSE TRUE
is.a.r(y ~ x)            # FALSE TRUE
is.a.r(expression(x+1)) # FALSE TRUE (not in 0.62.3!)
```

`is.single` *Is an Object of Single Precision Type?*

Description

`is.single` reports an error. There are no single precision values in R.

Usage

```
is.single(x)
```

Arguments

`x`	object to be tested.

Details

`is.single` is generic: you can write methods to handle specific classes of objects, see InternalMethods.

References

Becker, R. A., Chambers, J. M. and Wilks, A. R. (1988) *The New S Language*. Wadsworth & Brooks/Cole.

`kronecker` *Kronecker products on arrays*

Description

Computes the generalised kronecker product of two arrays, `X` and `Y`. `kronecker(X, Y)` returns an array `A` with dimensions `dim(X) * dim(Y)`.

Usage

```
kronecker(X, Y, FUN = "*", make.dimnames = FALSE, ...)
X %x% Y
```

Arguments

`X`	A vector or array.
`Y`	A vector or array.
`FUN`	a function; it may be a quoted string.
`make.dimnames`	Provide dimnames that are the product of the dimnames of `X` and `Y`.
`...`	optional arguments to be passed to `FUN`.

Details

If `X` and `Y` do not have the same number of dimensions, the smaller array is padded with dimensions of size one. The returned array comprises submatrices constructed by taking `X` one term at a time and expanding that term as `FUN(x, Y, ...)`.

`%x%` is an alias for `kronecker` (where `FUN` is hardwired to `"*"`).

Author(s)

Jonathan Rougier

References

Shayle R. Searle (1982) *Matrix Algebra Useful for Statistics.* John Wiley and Sons.

See Also

outer, on which kronecker is built and %*% for usual matrix multiplication.

Examples

```
# simple scalar multiplication
( M <- matrix(1:6, ncol=2) )
kronecker(4, M)
# Block diagonal matrix:
kronecker(diag(1, 3), M)

# ask for dimnames
fred <- matrix(1:12, 3, 4,
               dimnames=list(LETTERS[1:3], LETTERS[4:7]))
bill <- c("happy" = 100, "sad" = 1000)
kronecker(fred, bill, make.dimnames = TRUE)

bill <- outer(bill, c("cat"=3, "dog"=4))
kronecker(fred, bill, make.dimnames = TRUE)
```

`lapply` *Apply a Function over a List or Vector*

Description

`lapply` returns a list of the same length as `X`. Each element of which is the result of applying `FUN` to the corresponding element of `X`.

`sapply` is a "user-friendly" version of `lapply` also accepting vectors as `X`, and returning a vector or matrix with `dimnames` if appropriate.

`replicate` is a wrapper for the common use of `sapply` for repeated evaluation of an expression (which will usually involve random number generation).

Usage

```
lapply(X, FUN, ...)
sapply(X, FUN, ..., simplify = TRUE, USE.NAMES = TRUE)

replicate(n, expr, simplify = TRUE)
```

Arguments

`X`	list or vector to be used.
`FUN`	the function to be applied. In the case of functions like +, `%*%`, etc., the function name must be quoted.
`...`	optional arguments to `FUN`.
`simplify`	logical; should the result be simplified to a vector or matrix if possible?
`USE.NAMES`	logical; if `TRUE` and if `X` is character, use `X` as `names` for the result unless it had names already.
`n`	Number of replications.
`expr`	Expression to evaluate repeatedly.

References

Becker, R. A., Chambers, J. M. and Wilks, A. R. (1988) *The New S Language.* Wadsworth & Brooks/Cole.

See Also

`apply`, `tapply`.

Examples

```
x <- list(a = 1:10, beta = exp(-3:3),
          logic = c(TRUE,FALSE,FALSE,TRUE))
# compute the list mean for each list element
lapply(x,mean)
# median and quartiles for each list element
lapply(x, quantile, probs = 1:3/4)
sapply(x, quantile)
str(i39 <- sapply(3:9, seq)) # list of vectors
sapply(i39, fivenum)

hist(replicate(100, mean(rexp(10))))
```

`Last.value` *Value of Last Evaluated Expression*

Description

The value of the internal evaluation of a top-level R expression is always assigned to `.Last.value` (in `package:base`) before further processing (e.g., printing).

Usage

```
.Last.value
```

Details

The value of a top-level assignment *is* put in `.Last.value`, unlike S.

Do not assign to `.Last.value` in the workspace, because this will always mask the object of the same name in `package:base`.

See Also

`eval`

Examples

```
## These will not work correctly from example(), but they
## will in make check or if pasted in, as example() does
## not run them at the top level
gamma(1:15)  # think of some intensive calculation...
fac14 <- .Last.value # keep them

library("eda") # returns invisibly
.Last.value    # shows what library(.) above returned
```

length *Length of a Vector or List*

Description

Get or set the length of vectors (including lists).

Usage

```
length(x)
length(x) <- value
```

Arguments

`x`	a vector or list.
`value`	an integer.

Details

`length` is generic: you can write methods to handle specific classes of objects, see InternalMethods.

The replacement form can be used to reset the length of a vector. If a vector is shortened, extra values are discarded and when a vector is lengthened, it is padded out to its new length with `NA`s.

Value

The length of `x` as an `integer` of length 1, if `x` is (or can be coerced to) a vector or list. Otherwise, `length` returns `NA`.

References

Becker, R. A., Chambers, J. M. and Wilks, A. R. (1988) *The New S Language.* Wadsworth & Brooks/Cole.

See Also

`nchar` for counting the number of characters in character vectors.

Examples

```
length(diag(4)) # = 16 (4 x 4)
length(options()) # 12 or more
length(y ~ x1 + x2 + x3) # 3
length(expression(x, {y <- x^2; y+2}, x^y)) # 3
```

levels *Levels Attributes*

Description

`levels` provides access to the levels attribute of a variable. The first form returns the value of the levels of its argument and the second sets the attribute.

The assignment form (`"levels<-"`) of `levels` is a generic function and new methods can be written for it. The most important method is that for `factors`:

Usage

```
levels(x)
levels(x) <- value
```

Arguments

`x`	an object, for example a factor.
`value`	A valid value for `levels(x)`. For the default method, `NULL` or a character vector. For the `factor` method, a vector of character strings with length at least the number of levels of `x`, or a named list specifying how to rename the levels.

References

Becker, R. A., Chambers, J. M. and Wilks, A. R. (1988) *The New S Language*. Wadsworth & Brooks/Cole.

See Also

`nlevels`.

Examples

```
## assign individual levels
x <- gl(2, 4, 8)
levels(x)[1] <- "low"
levels(x)[2] <- "high"
x
```

```
## or as a group
y <- gl(2, 4, 8)
levels(y) <- c("low", "high")
y

## combine some levels
z <- gl(3, 2, 12)
levels(z) <- c("A", "B", "A")
z

## same, using a named list
z <- gl(3, 2, 12)
levels(z) <- list(A=c(1,3), B=2)
z

## we can add levels this way:
f <- factor(c("a","b"))
levels(f) <- c("c", "a", "b")
f

f <- factor(c("a","b"))
levels(f) <- list(C="C", A="a", B="b")
f
```

library *Loading and Listing of Packages*

Description

library and require load add-on packages.

.First.lib is called when a package is loaded; .Last.lib is called when a package is detached.

.packages returns information about package availability.

.path.package returns information about where a package was loaded from.

.find.package returns the directory paths of installed packages.

Usage

```
library(package, help, pos = 2, lib.loc = NULL,
        character.only = FALSE, logical.return = FALSE,
        warn.conflicts = TRUE,
        keep.source = getOption("keep.source.pkgs"),
        verbose = getOption("verbose"), version)
require(package, quietly = FALSE, warn.conflicts = TRUE,
        keep.source = getOption("keep.source.pkgs"),
        character.only = FALSE, version, save = TRUE)

.First.lib(libname, pkgname)
.Last.lib(libpath)

.packages(all.available = FALSE, lib.loc = NULL)
.path.package(package = .packages(), quiet = FALSE)
.find.package(package, lib.loc = NULL, quiet = FALSE,
              verbose = getOption("verbose"))
.libPaths(new)

.Library
.Autoloaded
```

Arguments

package, help the name of a package, given as a name or literal character string, or a character string, depending on whether character.only is FALSE (default) or TRUE).

`pos` the position on the search list at which to attach the loaded package. Note that `.First.lib` may attach other packages, and `pos` is computed *after* `.First.lib` has been run. Can also be the name of a position on the current search list as given by `search()`.

`lib.loc` a character vector describing the location of R library trees to search through, or `NULL`. The default value of `NULL` corresponds to all libraries currently known.

`character.only`
a logical indicating whether `package` or `help` can be assumed to be character strings.

`version` A character string denoting a version number of the package to be loaded. If no version is given, a suitable default is chosen.

`logical.return`
logical. If it is `TRUE`, `FALSE` or `TRUE` is returned to indicate success.

`warn.conflicts`
logical. If `TRUE`, warnings are printed about `conflicts` from attaching the new package, unless that package contains an object `.conflicts.OK`.

`keep.source` logical. If `TRUE`, functions "keep their source" including comments, see argument `keep.source` to `options`.

`verbose` a logical. If `TRUE`, additional diagnostics are printed.

`quietly` a logical. If `TRUE`, no message confirming package loading is printed.

`save` logical or environment. If `TRUE`, a call to `require` from the source for a package will save the name of the required package in the variable `".required"`, allowing function `detach` to warn if a required package is detached. See section 'Packages that require other packages' below.

`libname` a character string giving the library directory where the package was found.

`pkgname` a character string giving the name of the package.

`libpath` a character string giving the complete path to the package.

`all.available` logical; if `TRUE` return a character vector of all available packages in `lib.loc`.

`quiet`	logical. For `.path.package`, should this not give warnings or an error if the package(s) are not loaded? For `.find.package`, should this not give warnings or an error if the package(s) are not found?
`new`	a character vector with the locations of R library trees.

Details

`library(package)` and `require(package)` both load the package with name `package`. `require` is designed for use inside other functions; it returns `FALSE` and gives a warning (rather than an error as `library()` does) if the package does not exist. Both functions check and update the list of currently loaded packages and do not reload code that is already loaded.

For large packages, setting `keep.source = FALSE` may save quite a bit of memory.

If `library` is called with no `package` or `help` argument, it lists all available packages in the libraries specified by `lib.loc`, and returns the corresponding information in an object of class `"libraryIQR"`. The structure of this class may change in future versions. In earlier versions of R, only the names of all available packages were returned; use `.packages(all = TRUE)` for obtaining these. Note that `installed.packages()` returns even more information.

`library(help = somename)` computes basic information about the package `somename`, and returns this in an object of class `"packageInfo"`. The structure of this class may change in future versions.

`.First.lib` is called when a package is loaded by `library`. It is called with two arguments, the name of the library directory where the package was found (i.e., the corresponding element of `lib.loc`), and the name of the package (in that order). It is a good place to put calls to `library.dynam` which are needed when loading a package into this function (don't call `library.dynam` directly, as this will not work if the package is not installed in a "standard" location). `.First.lib` is invoked after the search path interrogated by `search()` has been updated, so `as.environment(match("package:name", search()))` will return the environment in which the package is stored. If calling `.First.lib` gives an error the loading of the package is abandoned, and the package will be unavailable. Similarly, if the option `".First.lib"` has a list element with the package's name, this element is called in the same manner as `.First.lib` when the package is loaded. This mechanism allows the user to set package "load hooks" in addition to startup code as provided by the package maintainers.

`.Last.lib` is called when a package is detached. Beware that it might be called if `.First.lib` has failed, so it should be written defensively. (It is called within `try`, so errors will not stop the package being detached.)

`.packages()` returns the "base names" of the currently attached packages *invisibly* whereas `.packages(all.available = TRUE)` gives (visibly) *all* packages available in the library location path `lib.loc`.

`.path.package` returns the paths from which the named packages were loaded, or if none were named, for all currently loaded packages. Unless `quiet = TRUE` it will warn if some of the packages named are not loaded, and given an error if none are. This function is not meant to be called by users, and its interface might change in future versions.

`.find.package` returns the paths to the locations where the given packages can be found. If `lib.loc` is `NULL`, then attached packages are searched before the libraries. If a package is found more than once, the first match is used. Unless `quiet = TRUE` a warning will be given about the named packages which are not found, and an error if none are. If `verbose` is true, warnings about packages found more than once are given. This function is not meant to be called by users, and its interface might change in future versions.

`.Autoloaded` contains the "base names" of the packages for which autoloading has been promised.

`.Library` is a character string giving the location of the default library, the '`library`' subdirectory of `R_HOME`. `.libPaths` is used for getting or setting the library trees that R knows about (and hence uses when looking for packages). If called with argument `new`, the library search path is set to the existing files in `unique(new, .Library)` and this is returned. If given no argument, a character vector with the currently known library trees is returned.

The library search path is initialized at startup from the environment variable `R_LIBS` (which should be a colon-separated list of directories at which R library trees are rooted) by calling `.libPaths` with the directories specified in `R_LIBS`.

Value

`library` returns the list of loaded (or available) packages (or `TRUE` if `logical.return` is `TRUE`). `require` returns a logical indicating whether the required package is available.

Packages that require other packages

The source code for a package that requires one or more other packages should have a call to `require`, preferably near the beginning of the source, and of course before any code that uses functions, classes or methods from the other package. The default for argument `save` will save the names of all required packages in the environment of the new package. The saved package names are used by `detach` when a package is detached to warn if other packages still require the detached package. Also, if a package is installed with saved image (see INSTALL), the saved package names are used to require these packages when the new package is attached.

Formal methods

`library` takes some further actions when package **methods** is attached (as it is by default). Packages may define formal generic functions as well as re-defining functions in other packages (notably **base**) to be generic, and this information is cached whenever such a package is loaded after **methods** and re-defined functions are excluded from the list of conflicts. The check requires looking for a pattern of objects; the pattern search may be avoided by defining an object `.noGenerics` (with any value) in the package. Naturally, if the package *does* have any such methods, this will prevent them from being used.

Note

`library` and `require` can only load an *installed* package, and this is detected by having a '`DESCRIPTION`' file containing a `Built:` field. Packages installed prior to 1.2.0 (released in December 2000) will need to be re-installed.

Under Unix-alikes, the code checks that the package was installed under a similar operating system as given by `R.version$platform` (the canonical name of the platform under which R was compiled), provided it contains compiled code. Packages which do not contain compiled code can be shared between Unix-alikes, but not to other OSes because of potential problems with line endings and OS-specific help files.

`library` and `require` use the underlying file system services to locate the libraries, with the result that on case-sensitive file systems package names are case-sensitive (i.e., `library(foo)` is different from `library(Foo)`), but they are not distinguished on case-insensitive file systems such as MS Windows. A warning is issued if the user specifies a name which isn't a perfect match to the package name, because future versions of R will require exact matches.

Author(s)

R core; Guido Masarotto for the `all.available=TRUE` part of `.packages`.

References

Becker, R. A., Chambers, J. M. and Wilks, A. R. (1988) *The New S Language*. Wadsworth & Brooks/Cole.

See Also

`attach`, `detach`, `search`, `objects`, `autoload`, `library.dynam`, `data`, `install.packages` and `installed.packages`; `INSTALL`, `REMOVE`.

Examples

```
(.packages())              # maybe just "base"
.packages(all = TRUE) # return all available as char vector
library()                  # list all available packages
library(lib = .Library)  # list packages in default library
library(help = eda)        # documentation on package 'eda'
library(eda)               # load package 'eda'
require(eda)               # the same
(.packages())              # "eda", too
detach("package:eda")

# if the package name is in a character vector, use
pkg <- "eda"
library(pkg, character.only = TRUE)
detach(pos = match(paste("package", pkg, sep=":"),
       search()))

require(pkg, character.only = TRUE)
detach(pos = match(paste("package", pkg, sep=":"),
       search()))

.path.package()
.Autoloaded   # maybe "ctest"

.libPaths()   # all library trees R knows about
```

```
require(nonexistent)        # FALSE

## Suppose a package needs to call a shared library named
## 'fooEXT', where 'EXT' is the system-specific extension.
## Then you should use
.First.lib <- function(lib, pkg) {
  library.dynam("foo", pkg, lib)
}

## if you want to mask as little as possible, use
library(mypkg, pos = "package:base")
```

library.dynam *Loading Shared Libraries*

Description

Load the specified file of compiled code if it has not been loaded already, or unloads it.

Usage

```
library.dynam(chname, package = .packages(), lib.loc = NULL,
              verbose = getOption("verbose"),
              file.ext = .Platform$dynlib.ext, ...)
library.dynam.unload(chname, libpath,
              verbose = getOption("verbose"),
              file.ext = .Platform$dynlib.ext)
.dynLibs(new)
```

Arguments

`chname`	a character string naming a shared library to load.
`package`	a character vector with the names of packages to search through.
`lib.loc`	a character vector describing the location of R library trees to search through, or `NULL`. The default value of `NULL` corresponds to all libraries currently known.
`libpath`	the path to the loaded package whose shared library is to be unloaded.
`verbose`	a logical value indicating whether an announcement is printed on the console before loading the shared library. The default value is taken from the verbose entry in the system options.
`file.ext`	the extension to append to the file name to specify the library to be loaded. This defaults to the appropriate value for the operating system.
`...`	additional arguments needed by some libraries that are passed to the call to `dyn.load` to control how the library is loaded.
`new`	a character vector of packages which have loaded shared libraries.

Details

`library.dynam` is designed to be used inside a package rather than at the command line, and should really only be used inside `.First.lib` on `.onLoad`. The system-specific extension for shared libraries (e.g., '`.so`' or '`.sl`' on Unix systems) should not be added.

`library.dynam.unload` is designed for use in `.Last.lib` or `.onUnload`.

`.dynLibs` is used for getting or setting the packages that have loaded shared libraries (using `library.dynam`). Versions of R prior to 1.6.0 used an internal global variable `.Dyn.libs` for storing this information: this variable is now defunct.

Value

`library.dynam` returns a character vector with the names of packages which have used it in the current R session to load shared libraries. This vector is returned as `invisible`, unless the `chname` argument is missing.

`library.dynam.unload` returns the updated character vector, invisibly.

References

Becker, R. A., Chambers, J. M. and Wilks, A. R. (1988) *The New S Language*. Wadsworth & Brooks/Cole.

See Also

`.First.lib`, `library`, `dyn.load`, `.packages`, `.libPaths`

`SHLIB` for how to create suitable shared libraries.

Examples

```
# which packages have been "dynamically loaded"
library.dynam()
```

`license` *The R License Terms*

Description

The license terms under which R is distributed.

Usage

```
license()
licence()
```

Details

R is distributed under the terms of the GNU GENERAL PUBLIC LICENSE Version 2, June 1991. A copy of this license is in '`$R_HOME/COPYING`'.

A small number of files (the API header files and import library) are distributed under the LESSER GNU GENERAL PUBLIC LICENSE version 2.1. A copy of this license is in '`$R_HOME/COPYING.LIB`'.

LINK *Create Executable Programs*

Description

Front-end for creating executable programs.

Usage

```
R CMD LINK [options] linkcmd
```

Arguments

`linkcmd`	a list of commands to link together suitable object files (include library objects) to create the executable program.
`options`	further options to control the linking, or for obtaining information about usage and version.

Details

The linker front-end is useful in particular when linking against the R shared library, in which case `linkcmd` must contain `-lR` but need not specify its library path.

Currently only works if the C compiler is used for linking, and no C++ code is used.

Use `R CMD LINK --help` for more usage information.

list   *Lists – Generic and Dotted Pairs*

Description

Functions to construct, coerce and check for all kinds of R lists.

Usage

```
list(...)
pairlist(...)

as.list(x, ...)
as.pairlist(x)

is.list(x)
is.pairlist(x)

alist(...)
```

Arguments

`...`	objects.
`x`	object to be coerced or tested.

Details

Most lists in R internally are *Generic Vectors*, whereas traditional *dotted pair* lists (as in LISP) are still available.

The arguments to `list` or `pairlist` are of the form `value` or `tag=value`. The functions return a list composed of its arguments with each value either tagged or untagged, depending on how the argument was specified.

`alist` is like `list`, except in the handling of tagged arguments with no value. These are handled as if they described function arguments with no default (cf. `formals`), whereas `list` simply ignores them.

`as.list` attempts to coerce its argument to list type. For functions, this returns the concatenation of the list of formals arguments and the function body. For expressions, the list of constituent calls is returned.

`is.list` returns `TRUE` if its argument is a `list` *or* a `pairlist` of `length`>0, whereas `is.pairlist` only returns `TRUE` in the latter case.

`is.list` and `is.pairlist` are generic: you can write methods to handle specific classes of objects, see InternalMethods.

An empty pairlist, `pairlist()` is the same as `NULL`. This is different from `list()`.

References

Becker, R. A., Chambers, J. M. and Wilks, A. R. (1988) *The New S Language*. Wadsworth & Brooks/Cole.

See Also

`vector(., mode="list")`, `c`, for concatenation; `formals`.

Examples

```
data(cars)
# create a plotting structure
pts <- list(x=cars[,1], y=cars[,2])
plot(pts)

# Argument lists
f <- function()x
# Note the specification of a "..." argument:
formals(f) <- al <- alist(x=, y=2, ...=)
f
str(al)

str(pl <- as.pairlist(ps.options()))

## These are all TRUE:
is.list(pl) && is.pairlist(pl)
!is.null(list())
is.null(pairlist())
!is.list(NULL)
is.pairlist(pairlist())
is.null(as.pairlist(list()))
is.null(as.pairlist(NULL))
```

list.files *List the Files in a Directory/Folder*

Description

This function produces a list containing the names of files in the named directory. `dir` is an alias.

Usage

```
list.files(path = ".", pattern = NULL, all.files = FALSE,
           full.names = FALSE, recursive = FALSE)
       dir(path = ".", pattern = NULL, all.files = FALSE,
           full.names = FALSE, recursive = FALSE)
```

Arguments

`path`	a character vector of full path names.
`pattern`	an optional regular expression. Only file names which match the regular expression will be returned.
`all.files`	a logical value. If `FALSE`, only the names of visible files are returned. If `TRUE`, all file names will be returned.
`full.names`	a logical value. If `TRUE`, the directory path is prepended to the file names. If `FALSE`, only the file names are returned.
`recursive`	logical. Should the listing recurse into directories?

Value

A character vector containing the names of the files in the specified directories, or `""` if there were no files. If a path does not exist or is not a directory or is unreadable it is skipped, with a warning.

The files are sorted in alphabetical order, on the full path if `full.names = TRUE`.

Note

File naming conventions are very platform dependent.

`recursive = TRUE` is not supported on all platforms, and may be ignored, with a warning.

Author(s)

Ross Ihaka, Brian Ripley

See Also

`file.info`, `file.access` and `files` for many more file handling functions.

Examples

```
list.files(R.home())
## Only files starting with a-l or r (*including*
## uppercase):
dir("../..", pattern = "^[a-lr]",full.names=TRUE)
```

load *Reload Saved Datasets*

Description

Reload the datasets written to a file with the function `save`.

Usage

```
load(file, envir = parent.frame())
loadURL(url, envir = parent.frame(), quiet = TRUE, ...)
```

Arguments

`file`	a connection or a character string giving the name of the file to load.
`envir`	the environment where the data should be loaded.
`url`	a character string naming a URL.
`quiet, ...`	additional arguments to `download.file`.

Details

`load` can load R objects saved in the current or any earlier format. It can read a compressed file (see `save`) directly from a file or from a suitable connection.

`loadURL` is a convenience wrapper which downloads a file, loads it and deletes the downloaded copy.

Value

A character vector of the names of objects created, invisibly.

See Also

`save`, `download.file`.

Examples

```
## save all data
save(list = ls(), file= "all.Rdata")

## restore the saved values to the current environment
load("all.Rdata")
```

```
## restore the saved values to the user's workspace
load("all.Rdata", .GlobalEnv)

## print the value to see what objects were created.
print(loadURL("http://host/file.sav"))
```

`localeconv` *Find Details of the Numerical Representations in the Current Locale*

Description

Get details of the numerical representations in the current locale.

Usage

```
Sys.localeconv()
```

Value

A character vector with 18 named components. See your ISO C documentation for details of the meaning.

It is possible to compile R without support for locales, in which case the value will be `NULL`.

See Also

`Sys.setlocale` for ways to set locales: by default R uses the C clocal for `"LC_NUMERIC"` and `"LC_MONETARY"`.

Examples

```
Sys.localeconv()
## The results in the default C locale are
#     decimal_point     thousands_sep          grouping
#               "."                ""                ""
#   int_curr_symbol   currency_symbol
#                ""                ""
# mon_decimal_point mon_thousands_sep      mon_grouping
#                ""                ""                ""
#     positive_sign     negative_sign   int_frac_digits
#                ""                ""             "127"
#       frac_digits      p_cs_precedes    p_sep_by_space
#             "127"              "127"             "127"
#      n_cs_precedes     n_sep_by_space
#             "127"              "127"
#       p_sign_posn        n_sign_posn
#             "127"              "127"
```

```
## Now try your default locale (which might be "C").
old <- Sys.getlocale()
Sys.setlocale(locale = "")
Sys.localeconv()
Sys.setlocale(locale = old)

read.table("foo", dec=Sys.localeconv()["decimal_point"])
```

locales *Query or Set Aspects of the Locale*

Description

Get details of or set aspects of the locale for the R process.

Usage

```
Sys.getlocale(category = "LC_ALL")
Sys.setlocale(category = "LC_ALL", locale = "")
```

Arguments

`category`	character string. Must be one of `"LC_ALL"`, `"LC_COLLATE"`, `"LC_CTYPE"`, `"LC_MONETARY"`, `"LC_NUMERIC"` or `"LC_TIME"`.
`locale`	character string. A valid locale name on the system in use. Normally `""` (the default) will pick up the default locale for the system.

Details

The locale describes aspects of the internationalization of a program. Initially most aspects of the locale of R are set to `"C"` (which is the default for the C language and reflects North-American usage). R does set `"LC_CTYPE"` and `"LC_COLLATE"`, which allow the use of a different character set (typically ISO Latin 1) and alphabetic comparisons in that character set (including the use of `sort`) and `"LC_TIME"` may affect the behaviour of `as.POSIXlt` and `strptime` and functions which use them (but not `date`).

R can be built with no support for locales, but it is normally available on Unix and is available on Windows.

Some systems will have other locale categories, but the six described here are those specified by POSIX.

Value

A character string of length one describing the locale in use (after setting for `Sys.setlocale`), or an empty character string if the locale is invalid (with a warning) or `NULL` if locale information is unavailable.

For category = "LC_ALL" the details of the string are system-specific: it might be a single locale or a set of locales separated by "/" (Solaris) or ";" (Windows). For portability, it is best to query categories individually. It is guaranteed that the result of foo <- Sys.getlocale() can be used in Sys.setlocale("LC_ALL", locale = foo) on the same machine.

Warning

Setting "LC_NUMERIC" can produce output that R cannot then read by scan or read.table with their default arguments, which are not locale-specific.

See Also

strptime for uses of category = "LC_TIME". Sys.localeconv for details of numerical representations.

Examples

```
Sys.getlocale()
Sys.getlocale("LC_TIME")

# Solaris: details are OS-dependent
Sys.setlocale("LC_TIME", "de")
# Windows
Sys.setlocale("LC_TIME", "German")
# turn off locale-specific sorting
Sys.setlocale("LC_COLLATE", "C")
```

Logic *Logical Operators*

Description

These operators act on logical vectors.

Usage

```
! x
x & y
x && y
x | y
x || y
xor(x, y)
```

Arguments

`x, y`	logical vectors

Details

`!` indicates logical negation (NOT).

`&` and `&&` indicate logical AND and `|` and `||` indicate logical OR. The shorter form performs elementwise comparisons in much the same way as arithmetic operators. The longer form evaluates left to right examining only the first element of each vector. Evaluation proceeds only until the result is determined. The longer form is appropriate for programming control-flow and typically preferred in `if` clauses.

`xor` indicates elementwise exclusive OR.

`NA` is a valid logical object. Where a component of `x` or `y` is `NA`, the result will be `NA` if the outcome is ambiguous. In other words `NA & TRUE` evaluates to `NA`, but `NA & FALSE` evaluates to `FALSE`. See the examples below.

References

Becker, R. A., Chambers, J. M. and Wilks, A. R. (1988) *The New S Language*. Wadsworth & Brooks/Cole.

See Also

TRUE or logical.

Syntax for operator precedence.

Examples

```
y <- 1 + (x <- rpois(50, lambda=1.5) / 4 - 1)
x[(x > 0) & (x < 1)]    # all x values between 0 and 1
if (any(x == 0) || any(y == 0)) "zero encountered"

## construct truth tables :
x <- c(NA, FALSE, TRUE)
names(x) <- as.character(x)
outer(x, x, "&") ## AND table
outer(x, x, "|") ## OR table
```

`logical` *Logical Vectors*

Description

Create or test for objects of type `"logical"`, and the basic logical "constants".

Usage

```
TRUE
FALSE
T; F

logical(length = 0)
as.logical(x, ...)
is.logical(x)
```

Arguments

`length`	desired length.
`x`	object to be coerced or tested.
`...`	further arguments passed to or from other methods.

Details

`TRUE` and `FALSE` are part of the R language, where `T` and `F` are global variables set to these. All four are `logical(1)` vectors.

`is.logical` is generic: you can write methods to handle specific classes of objects, see InternalMethods.

Value

`logical` creates a logical vector of the specified length. Each element of the vector is equal to `FALSE`.

`as.logical` attempts to coerce its argument to be of logical type. For `factors`, this uses the `levels` (labels) and not the `codes`. Like `as.vector` it strips attributes including names.

`is.logical` returns `TRUE` or `FALSE` depending on whether its argument is of logical type or not.

References

Becker, R. A., Chambers, J. M. and Wilks, A. R. (1988) *The New S Language*. Wadsworth & Brooks/Cole.

`lower.tri` *Lower and Upper Triangular Part of a Matrix*

Description

Returns a matrix of logicals the same size of a given matrix with entries `TRUE` in the lower or upper triangle.

Usage

```
lower.tri(x, diag = FALSE)
upper.tri(x, diag = FALSE)
```

Arguments

`x`	a matrix.
`diag`	logical. Should the diagonal be included?

See Also

`diag`, `matrix`.

Examples

```
(m2 <- matrix(1:20, 4, 5))
lower.tri(m2)
m2[lower.tri(m2)] <- NA
m2
```

lowess *Scatter Plot Smoothing*

Description

This function performs the computations for the *LOWESS* smoother (see the reference below). `lowess` returns a list containing components `x` and `y` which give the coordinates of the smooth. The smooth should be added to a plot of the original points with the function `lines`.

Usage

```
lowess(x, y = NULL, f = 2/3, iter=3,
       delta = 0.01 * diff(range(xy$x[o])))
```

Arguments

`x, y`	vectors giving the coordinates of the points in the scatter plot. Alternatively a single plotting structure can be specified.
`f`	the smoother span. This gives the proportion of points in the plot which influence the smooth at each value. Larger values give more smoothness.
`iter`	the number of robustifying iterations which should be performed. Using smaller values of `iter` will make `lowess` run faster.
`delta`	values of `x` which lie within `delta` of each other are replaced by a single value in the output from `lowess`. Defaults to 1/100th of the range of `x`.

References

Becker, R. A., Chambers, J. M. and Wilks, A. R. (1988) *The New S Language.* Wadsworth & Brooks/Cole.

Cleveland, W. S. (1979) Robust locally weighted regression and smoothing scatterplots. *J. Amer. Statist. Assoc.* **74**, 829–836.

Cleveland, W. S. (1981) LOWESS: A program for smoothing scatterplots by robust locally weighted regression. *The American Statistician*, **35**, 54.

See Also

`loess` (in package **modreg**), a newer formula based version of `lowess` (with different defaults!).

Examples

```
data(cars)
plot(cars, main = "lowess(cars)")
lines(lowess(cars), col = 2)
lines(lowess(cars, f=.2), col = 3)
legend(5, 120, c(paste("f = ", c("2/3", ".2"))),
       lty = 1, col = 2:3)
```

ls *List Objects*

Description

`ls` and `objects` return a vector of character strings giving the names of the objects in the specified environment. When invoked with no argument at the top level prompt, `ls` shows what data sets and functions a user has defined. When invoked with no argument inside a function, `ls` returns the names of the functions local variables. This is useful in conjunction with `browser`.

Usage

```
ls(name, pos = -1, envir = as.environment(pos),
   all.names = FALSE, pattern)
objects(name, pos= -1, envir = as.environment(pos),
        all.names = FALSE, pattern)
```

Arguments

`name`	which environment to use in listing the available objects. Defaults to the *current* environment. Although called `name` for back compatibility, in fact this argument can specify the environment in any form; see the details section.
`pos`	An alternative argument to `name` for specifying the environment as a position in the search list. Mostly there for back compatibility.
`envir`	an alternative argument to `name` for specifying the environment evaluation environment. Mostly there for back compatibility.
`all.names`	a logical value. If `TRUE`, all object names are returned. If `FALSE`, names which begin with a '`.`' are omitted.
`pattern`	an optional regular expression. Only names matching `pattern` are returned.

Details

The `name` argument can specify the environment from which object names are taken in one of several forms: as an integer (the position in the `search` list); as the character string name of an element in the search list; or as an explicit `environment` (including using `sys.frame` to access the currently active function calls). By default, the environment of the call to `ls` or `objects` is used. The `pos` and `envir` arguments are an alternative way to specify an environment, but are primarily there for back compatibility.

References

Becker, R. A., Chambers, J. M. and Wilks, A. R. (1988) *The New S Language.* Wadsworth & Brooks/Cole.

See Also

`apropos` (or `find`) for finding objects in the whole search path; `grep` for more details on "regular expressions"; `class`, `methods`, etc., for object-oriented programming.

Examples

```
.Ob <- 1
ls(pat="O")
ls(pat="O", all = TRUE)    # also shows ".[foo]"

# shows an empty list because inside myfunc no variables
# are defined
myfunc <- function() {ls()}
myfunc()

# define a local variable inside myfunc
myfunc <- function() {y <- 1; ls()}
myfunc()                   # shows "y"
```

mad *Median Absolute Deviation*

Description

Compute the median absolute deviation, i.e., the (lo-/hi-) median of the absolute deviations from the median, and (by default) adjust by a factor for asymptotically normal consistency.

Usage

```
mad(x, center = median(x), constant = 1.4826, na.rm = FALSE,
    low = FALSE, high = FALSE)
```

Arguments

`x`	a numeric vector.
`center`	Optionally, the centre: defaults to the median.
`constant`	scale factor.
`na.rm`	if `TRUE` then `NA` values are stripped from `x` before computation takes place.
`low`	if `TRUE`, compute the "lo-median", i.e., for even sample size, do not average the two middle values, but take the smaller one.
`high`	if `TRUE`, compute the "hi-median", i.e., take the larger of the two middle values for even sample size.

Details

The actual value calculated is `constant * cMedian(abs(x - center))` with the default value of `center` being `median(x)`, and `cMedian` being the usual, the "low" or "high" median, see the arguments description for `low` and `high` above.

The default `constant = 1.4826` (approximately $1/\Phi^{-1}(\frac{3}{4})$ = `1/qnorm(3/4)`) ensures consistency, i.e.,

$$E[mad(X_1, \ldots, X_n)] = \sigma$$

for X_i distributed as $N(\mu, \sigma^2)$ and large n.

If `na.rm` is `TRUE` then `NA` values are stripped from `x` before computation takes place. If this is not done then an `NA` value in `x` will cause `mad` to return `NA`.

See Also

IQR which is simpler but less robust, **median**, **var**.

Examples

```
mad(c(1:9))
print(mad(c(1:9),     constant=1)) ==
      mad(c(1:8,100), constant=1)       # = 2 ; TRUE
x <- c(1,2,3, 5,7,8)
sort(abs(x - median(x)))
c(mad(x, co=1), mad(x, co=1, lo = TRUE),
  mad(x, co=1, hi = TRUE))
```

mahalanobis *Mahalanobis Distance*

Description

Returns the Mahalanobis distance of all rows in `x` and the vector μ =`center` with respect to Σ =`cov`. This is (for vector `x`) defined as

$$D^2 = (x - \mu)'\Sigma^{-1}(x - \mu)$$

Usage

```
mahalanobis(x, center, cov, inverted=FALSE, tol.inv = 1e-7)
```

Arguments

`x`	vector or matrix of data with, say, p columns.
`center`	mean vector of the distribution or second data vector of length p.
`cov`	covariance matrix ($p \times p$) of the distribution.
`inverted`	logical. If `TRUE`, `cov` is supposed to contain the *inverse* of the covariance matrix.
`tol.inv`	tolerance to be used for computing the inverse (if `inverted` is false), see `solve`.

Author(s)

Friedrich Leisch

See Also

`cov`, `var`

Examples

```
ma <- cbind(1:6, 1:3)
(S <-  var(ma))
mahalanobis(c(0,0), 1:2, S)

x <- matrix(rnorm(100*3), ncol = 3)
stopifnot(mahalanobis(x, 0, diag(ncol(x))) == rowSums(x*x))
# Here, D^2 = usual Euclidean distances
```

```
Sx <- cov(x)
D2 <- mahalanobis(x, rowMeans(x), Sx)
plot(density(D2, bw=.5), main="Mahalanobis distances,
     n=100, p=3"); rug(D2)
qqplot(qchisq(ppoints(100), df=3), D2,
       main = expression("Q-Q plot of Mahalanobis" * ~D^2 *
                         " vs. quantiles of" * ~ chi[3]^2))
abline(0, 1, col = 'gray')
```

make.names *Make Syntactically Valid Names*

Description

Make syntactically valid names out of character vectors.

Usage

```
make.names(names, unique = FALSE)
```

Arguments

`names`	character vector to be coerced to syntactically valid names. This is coerced to character if necessary.
`unique`	logical; if `TRUE`, the resulting elements are unique. This may be desired for, e.g., column names.

Details

A syntactically valid name consists of letters, numbers, and the dot character and starts with a letter or the dot. Names such as `".2"` are not valid, and neither are the reserved words.

The character `"X"` is prepended if necessary. All invalid characters are translated to `"."`. A missing value is translated to `"NA"`. Names which match R keywords have a dot appended to them. Duplicated values are altered by `make.unique`.

Value

A character vector of same length as `names` with each changed to a syntactically valid name.

See Also

`make.unique`, `names`, `character`, `data.frame`.

Examples

```
make.names(c("a and b", "a_and_b"), unique=TRUE)
# "a.and.b"  "a.and.b.1"

data(state)
```

```
# those 10 with a space
state.name[make.names(state.name) != state.name]
```

make.packages.html *Update HTML documentation files*

Description

Functions to re-create the HTML documentation files to reflect all installed packages.

Usage

```
make.packages.html(lib.loc = .libPaths())
```

Arguments

`lib.loc`	character vector. List of libraries to be included.

Details

This sets up the links from packages in libraries to the '`.R`' subdirectory of the per-session directory (see `tempdir`) and then creates the '`packages.html`' and '`index.txt`' files to point to those links.

If a package is available in more than one library tree, all the copies are linked, after the first with suffix `.1` etc.

Value

Logical, whether the function succeeded in recreating the files.

See Also

`help.start`

`make.socket` *Create a Socket Connection*

Description

With `server = FALSE` attempts to open a client socket to the specified port and host. With `server = TRUE` listens on the specified port for a connection and then returns a server socket. It is a good idea to use `on.exit` to ensure that a socket is closed, as you only get 64 of them.

Usage

```
make.socket(host = "localhost", port, fail = TRUE,
            server = FALSE)
```

Arguments

`host`	name of remote host
`port`	port to connect to/listen on
`fail`	failure to connect is an error?
`server`	a server socket?

Value

An object of class `"socket"`.

`socket`	socket number. This is for internal use
`port`	port number of the connection
`host`	name of remote computer

Warning

I don't know if the connecting host name returned when `server = TRUE` can be trusted. I suspect not.

Author(s)

Thomas Lumley

References

Adapted from Luke Tierney's code for `XLISP-Stat`, in turn based on code from Robbins and Robbins "Practical UNIX Programming"

See Also

`close.socket`, `read.socket`

Examples

```
daytime <- function(host = "localhost"){
    a <- make.socket(host, 13)
    on.exit(close.socket(a))
    read.socket(a)
}
## Offical time (UTC) from US Naval Observatory
daytime("tick.usno.navy.mil")
```

`make.tables` *Create model.tables*

Description

These are support functions for (the methods of) `model.tables` and probably not much of use otherwise.

Usage

```
make.tables.aovproj    (proj.cols, mf.cols, prjs, mf,
                        fun = "mean", prt = FALSE, ...)

make.tables.aovprojlist(proj.cols, strata.cols, model.cols,
                        projections, model, eff,
                        fun = "mean", prt = FALSE, ...)
```

See Also

`model.tables`

make.unique *Make Character Strings Unique*

Description

Makes the elements of a character vector unique by appending sequence numbers to duplicates.

Usage

```
make.unique(names, sep = ".")
```

Arguments

names	a character vector
sep	a character string used to separate a duplicate name from its sequence number.

Details

The algorithm used by `make.unique` has the property that `make.unique(c(A, B)) == make.unique(c(make.unique(A), B))`.

In other words, you can append one string at a time to a vector, making it unique each time, and get the same result as applying `make.unique` to all of the strings at once.

If character vector `A` is already unique, then `make.unique(c(A, B))` preserves `A`.

Value

A character vector of same length as `names` with duplicates changed.

Author(s)

Thomas P Minka

See Also

`make.names`

Examples

```
make.unique(c("a", "a", "a"))
make.unique(c(make.unique(c("a", "a")), "a"))

make.unique(c("a", "a", "a.2", "a"))
make.unique(c(make.unique(c("a", "a")), "a.2", "a"))

rbind(data.frame(x=1), data.frame(x=2), data.frame(x=3))
rbind(rbind(data.frame(x=1), data.frame(x=2)),
      data.frame(x=3))
```

`manglePackageName` *Mangle the Package Name*

Description

This function takes the package name and the package version number and pastes them together with a separating underscore.

Usage

```
manglePackageName(pkgName, pkgVersion)
```

Arguments

`pkgName`	The package name, as a character string.
`pkgVersion`	The package version, as a character string.

Value

A character string with the two inputs pasted together.

Examples

```
manglePackageName("foo", "1.2.3")
```

mapply *Apply a function to multiple list or vector arguments*

Description

A multivariate version of `sapply`. `mapply` applies `FUN` to the first elements of each `...` argument, the second elements, the third elements, and so on. Arguments are recycled if necessary.

Usage

```
mapply(FUN, ..., MoreArgs = NULL, SIMPLIFY = TRUE,
       USE.NAMES = TRUE)
```

Arguments

`FUN`	Function to apply
`...`	Arguments to vectorise over (list or vector)
`MoreArgs`	A list of other arguments to `FUN`
`SIMPLIFY`	Attempt to reduce the result to a vector or matrix?
`USE.NAMES`	If the first `...` argument is character and the result doesn't already have names, use it as the names

Value

A list, vector, or matrix.

See Also

`sapply`

Examples

```
mapply(rep, 1:4, 4:1)

mapply(rep, times=1:4, x=4:1)

mapply(rep, times=1:4, MoreArgs=list(x=42))
```

`margin.table` *Compute table margin*

Description

For a contingency table in array form, compute the sum of table entries for a given index.

Usage

```
margin.table(x, margin=NULL)
```

Arguments

`x`	an array
`margin`	index number (1 for rows, etc.)

Details

This is really just `apply(x, margin, sum)` packaged up for newbies, except that if `margin` has length zero you get `sum(x)`.

Value

The relevant marginal table. The class of `x` is copied to the output table, except in the summation case.

Author(s)

Peter Dalgaard

Examples

```
m<-matrix(1:4,2)
margin.table(m,1)
margin.table(m,2)
```

`mat.or.vec` *Create a Matrix or a Vector*

Description

`mat.or.vec` creates an `nr` by `nc` zero matrix if `nc` is greater than 1, and a zero vector of length `nr` if `nc` equals 1.

Usage

```
mat.or.vec(nr, nc)
```

Arguments

`nr, nc`	numbers of rows and columns.

Examples

```
mat.or.vec(3, 1)
mat.or.vec(3, 2)
```

`match` *Value Matching*

Description

`match` returns a vector of the positions of (first) matches of its first argument in its second.

`%in%` is a more intuitive interface as a binary operator, which returns a logical vector indicating if there is a match or not for its left operand.

Usage

```
match(x, table, nomatch = NA, incomparables = FALSE)

x %in% table
```

Arguments

`x`	the values to be matched.
`table`	the values to be matched against.
`nomatch`	the value to be returned in the case when no match is found. Note that it is coerced to `integer`.
`incomparables`	a vector of values that cannot be matched. Any value in `x` matching a value in this vector is assigned the `nomatch` value. Currently, `FALSE` is the only possible value, meaning that all values can be matched.

Details

`%in%` is currently defined as `"%in%" <- function(x,table) match(x,table,nomatch=0) > 0`

Factors are converted to character vectors, and then `x` and `table` are coerced to a common type (the later of the two types in R's ordering, logical $<$ integer $<$ numeric $<$ complex $<$ character) before matching.

Value

In both cases, a vector of the same length as `x`.

`match`: An integer vector giving the position in `table` of the first match if there is a match, otherwise `nomatch`.

If `x[i]` is found to equal `table[j]` then the value returned in the `i`-th position of the return value is `j`, for the smallest possible `j`. If no match is found, the value is `nomatch`.

`%in%`: A logical vector, indicating if a match was located for each element of `x`.

References

Becker, R. A., Chambers, J. M. and Wilks, A. R. (1988) *The New S Language.* Wadsworth & Brooks/Cole.

See Also

`pmatch` and `charmatch` for (*partial*) string matching, `match.arg`, etc for function argument matching.

`is.element` for an S-compatible equivalent of `%in%`.

Examples

```
## The intersection of two sets :
intersect <- function(x, y) y[match(x, y, nomatch = 0)]
intersect(1:10,7:20)

1:10 %in% c(1,3,5,9)
sstr <- c("c","ab","B","bba","c","@","bla","a","Ba","%")
sstr[sstr %in% c(letters,LETTERS)]

"%w/o%" <- function(x,y) x[!x %in% y] # x without y
(1:10) %w/o% c(3,7,12)
```

match.arg *Argument Verification Using Partial Matching*

Description

`match.arg` matches `arg` against a table of candidate values as specified by `choices`.

Usage

```
match.arg(arg, choices)
```

Arguments

`arg`	a character string
`choices`	a character vector of candidate values

Details

In the one-argument form `match.arg(arg)`, the choices are obtained from a default setting for the formal argument `arg` of the function from which `match.arg` was called.

Matching is done using `pmatch`, so `arg` may be abbreviated.

Value

The unabbreviated version of the unique partial match if there is one; otherwise, an error is signalled.

See Also

`pmatch`, `match.fun`, `match.call`.

Examples

```
## Extends the example for 'switch'
center <- function(x, type = c("mean","median","trimmed")) {
  type <- match.arg(type)
  switch(type,
         mean = mean(x),
         median = median(x),
         trimmed = mean(x, trim = .1))
}
```

```
x <- rcauchy(10)
center(x, "t")       # Works
center(x, "med")     # Works
center(x, "m")       # Error
```

match.call *Argument Matching*

Description

`match.call` returns a call in which all of the arguments are specified by their names. The most common use is to get the call of the current function, with all arguments named.

Usage

```
match.call(definition = NULL, call = sys.call(sys.parent()),
           expand.dots = TRUE)
```

Arguments

`definition`	a function, by default the function from which `match.call` is called.
`call`	an unevaluated call to the function specified by `definition`, as generated by `call`.
`expand.dots`	logical. Should arguments matching `...` in the call be included or left as a `...` argument?

Value

An object of class `call`.

References

Chambers, J. M. (1998) *Programming with Data. A Guide to the S Language.* Springer.

See Also

`call`, `pmatch`, `match.arg`, `match.fun`.

Examples

```
match.call(get, call("get", "abc", i = FALSE, p = 3))
## -> get(x = "abc", pos = 3, inherits = FALSE)
fun <- function(x, lower = 0, upper = 1) {
  structure((x - lower) / (upper - lower),
            CALL = match.call())
```

```
}
fun(4 * atan(1), u = pi)
```

`match.fun` *Function Verification for "Function Variables"*

Description

When called inside functions that take a function as argument, extract the desired function object while avoiding undesired matching to objects of other types.

Usage

```
match.fun(FUN, descend = TRUE)
```

Arguments

`FUN`	item to match as function.
`descend`	logical; control whether to search past non-function objects.

Details

`match.fun` is not intended to be used at the top level since it will perform matching in the *parent* of the caller.

If `FUN` is a function, it is returned. If it is a symbol or a character vector of length one, it will be looked up using `get` in the environment of the parent of the caller. If it is of any other mode, it is attempted first to get the argument to the caller as a symbol (using `substitute` twice), and if that fails, an error is declared.

If `descend = TRUE`, `match.fun` will look past non-function objects with the given name; otherwise if `FUN` points to a non-function object then an error is generated.

This is now used in base functions such as `apply`, `lapply`, `outer`, and `sweep`.

Value

A function matching `FUN` or an error is generated.

Bugs

The `descend` argument is a bit of misnomer and probably not actually needed by anything. It may go away in the future.

It is impossible to fully foolproof this. If one `attaches` a list or data frame containing a character object with the same name of a system function, it will be used.

Author(s)

Peter Dalgaard and Robert Gentleman, based on an earlier version by Jonathan Rougier.

See Also

`match.arg`, `get`

Examples

```
# Same as get("*"):
match.fun("*")
# Overwrite outer with a vector
outer <- 1:5

match.fun(outer, descend = FALSE) # Error: not a function

match.fun(outer) # finds it anyway
is.function(match.fun("outer")) # as well
```

maxCol *Find Maximum Position in Matrix*

Description

Find the maximum position for each row of a matrix, breaking ties at random.

Usage

```
max.col(m)
```

Arguments

m	numerical matrix

Details

Ties are broken at random. The determination of "tie" assumes that the entries are probabilities: there is a relative tolerance of 10^{-5}, relative to the largest entry in the row.

Value

index of a maximal value for each row, an integer vector of length `nrow(m)`.

References

Venables, W. N. and Ripley, B. D. (2002) *Modern Applied Statistics with S.* New York: Springer (4th ed).

See Also

`which.max` for vectors.

Examples

```
data(swiss)
# mostly "1" and "5", 5 x "2" and once "4"
table(mc <- max.col(swiss))
# 3 33 45 45 33 6
swiss[unique(print(mr <- max.col(t(swiss)))) , ]
```

mean *Arithmetic Mean*

Description

Generic function for the (trimmed) arithmetic mean.

Usage

```
mean(x, ...)

## Default S3 method:
mean(x, trim = 0, na.rm = FALSE, ...)
```

Arguments

`x`	An R object. Currently there are methods for numeric data frames, numeric vectors and dates. A complex vector is allowed for `trim = 0`, only.
`trim`	the fraction (0 to 0.5) of observations to be trimmed from each end of `x` before the mean is computed.
`na.rm`	a logical value indicating whether `NA` values should be stripped before the computation proceeds.
`...`	further arguments passed to or from other methods.

Value

For a data frame, a named vector with the appropriate method being applied column by column.

If `trim` is zero (the default), the arithmetic mean of the values in `x` is computed.

If `trim` is non-zero, a symmetrically trimmed mean is computed with a fraction of `trim` observations deleted from each end before the mean is computed.

References

Becker, R. A., Chambers, J. M. and Wilks, A. R. (1988) *The New S Language*. Wadsworth & Brooks/Cole.

See Also

`weighted.mean`, `mean.POSIXct`

Examples

```
x <- c(0:10, 50)
xm <- mean(x)
c(xm, mean(x, trim = 0.10))

data(USArrests)
mean(USArrests, trim = 0.2)
```

`median` *Median Value*

Description

Compute the sample median of the vector of values given as its argument.

Usage

```
median(x, na.rm=FALSE)
```

Arguments

`x`	a numeric vector containing the values whose median is to be computed.
`na.rm`	a logical value indicating whether `NA` values should be stripped before the computation proceeds.

References

Becker, R. A., Chambers, J. M. and Wilks, A. R. (1988) *The New S Language.* Wadsworth & Brooks/Cole.

See Also

`quantile` for general quantiles.

Examples

```
median(1:4) # = 2.5 [even number]
median(c(1:3,100,1000)) # = 3 [odd, robust]
```

Memory *Memory Available for Data Storage*

Description

Use command line options to control the memory available for R.

Usage

```
R --min-vsize=vl --max-vsize=vu --min-nsize=nl --max-nsize=nu

mem.limits(nsize = NA, vsize = NA)
```

Arguments

`vl, vu, vsize`	Heap memory in bytes.
`nl, nu, nsize`	Number of cons cells.

Details

R has a variable-sized workspace (from version 1.2.0). There is now much less need to set memory options than previously, and most users will never need to set these. They are provided both as a way to control the overall memory usage (which can also be done by operating-system facilities such as `limit` on Unix), and since setting larger values of the minima will make R slightly more efficient on large tasks.

To understand the options, one needs to know that R maintains separate areas for fixed and variable sized objects. The first of these is allocated as an array of "*cons cells*" (Lisp programmers will know what they are, others may think of them as the building blocks of the language itself, parse trees, etc.), and the second are thrown on a "*heap*" of "Vcells" of 8 bytes each. Effectively, the input `v` is rounded up to the nearest multiple of 8.

Each cons cell occupies 28 bytes on a 32-bit machine, (usually) 56 bytes on a 64-bit machine.

The '`--*-nsize`' options can be used to specify the number of cons cells and the '`--*-vsize`' options specify the size of the vector heap in bytes. Both options must be integers or integers followed by `G`, `M`, `K`, or `k` meaning `Giga` ($2^{30} = 1073741824$) *Mega* ($2^{20} = 1048576$), (computer) *Kilo* ($2^{10} = 1024$), or regular *kilo* (1000).

The '--min-*' options set the minimal sizes for the number of cons cells and for the vector heap. These values are also the initial values, but thereafter R will grow or shrink the areas depending on usage, but never exceeding the limits set by the '--max-*' options nor decreasing below the initial values.

The default values are currently minima of 350k cons cells, 6Mb of vector heap and no maxima (other than machine resources). The maxima can be changed during an R session by calling `mem.limits`. (If this is called with the default values, it reports the current settings.)

You can find out the current memory consumption (the heap and cons cells used as numbers and megabytes) by typing `gc()` at the R prompt. Note that following `gcinfo(TRUE)`, automatic garbage collection always prints memory use statistics. Maxima will never be reduced below the current values for triggering garbage collection, and attempts to do so will be silently ignored.

When using `read.table`, the memory requirements are in fact higher than anticipated, because the file is first read in as one long string which is then split again. Use `scan` if possible in case you run out of memory when reading in a large table.

Value

(`mem.limits`) an integer vector giving the current settings of the maxima, possibly `NA`.

Note

For backwards compatibility, options '--nsize' and '--vsize' are equivalent to '--min-nsize' and '--min-vsize'.

See Also

`gc` for information on the garbage collector, `memory.profile` for profiling the usage of cons cells.

Examples

```
# Start R with 10MB of heap memory and 500k cons cells,
# limit to 100Mb and 1M cells

## Unix
R --min-vsize=10M --max-vsize=100M --min-nsize=500k
 --max-nsize=1M
```

memory.profile *Profile the Usage of Cons Cells*

Description

Lists the usage of the cons cells by SEXPREC type.

Usage

```
memory.profile()
```

Details

The current types and their uses are listed in the include file 'Rinternals.h'. There will be blanks in the list corresponding to types that are no longer in use (types 11 and 12 at the time of writing). Also FUNSXP is not included.

Value

A vector of counts, named by the types.

See Also

gc for the overall usage of cons cells.

Examples

```
memory.profile()
```

`menu` *Menu Interaction Function*

Description

`menu` presents the user with a menu of choices labelled from 1 to the number of choices. To exit without choosing an item one can select '`0`'.

Usage

```
menu(choices, graphics = FALSE, title = "")
```

Arguments

`choices`	a character vector of choices
`graphics`	a logical indicating whether a graphics menu should be used. Currently unused.
`title`	a character string to be used as the title of the menu

Value

The number corresponding to the selected item, or 0 if no choice was made.

References

Becker, R. A., Chambers, J. M. and Wilks, A. R. (1988) *The New S Language.* Wadsworth & Brooks/Cole.

Examples

```
switch(menu(c("List letters", "List LETTERS")) + 1,
       cat("Nothing done\n"), letters, LETTERS)
```

merge *Merge Two Data Frames*

Description

Merge two data frames by common columns or row names, or do other versions of database "join" operations.

Usage

```
merge(x, y, ...)

## Default S3 method:
merge(x, y, ...)

## S3 method for class 'data.frame':
merge(x, y, by = intersect(names(x), names(y)),
      by.x = by, by.y = by, all = FALSE,
      all.x = all, all.y = all,
      sort = TRUE, suffixes = c(".x",".y"), ...)
```

Arguments

`x, y`	data frames, or objects to be coerced to one
`by, by.x, by.y`	specifications of the common columns. See Details.
`all`	logical; `all=L` is shorthand for `all.x=L` and `all.y=L`.
`all.x`	logical; if `TRUE`, then extra rows will be added to the output, one for each row in `x` that has no matching row in `y`. These rows will have `NA`s in those columns that are usually filled with values from `y`. The default is `FALSE`, so that only rows with data from both `x` and `y` are included in the output.
`all.y`	logical; analogous to `all.x` above.
`sort`	logical. Should the results be sorted on the `by` columns?
`suffixes`	character(2) specifying the suffixes to be used for making non-`by` `names()` unique.
`...`	arguments to be passed to or from methods.

Details

By default the data frames are merged on the columns with names they both have, but separate specifications of the columns can be given by `by.x` and `by.y`. Columns can be specified by name, number or by a logical vector: the name `"row.names"` or the number `0` specifies the row names. The rows in the two data frames that match on the specified columns are extracted, and joined together. If there is more than one match, all possible matches contribute one row each.
If the `by.*` vectors are of length 0, the result, `r`, is the "Cartesian product" of `x` and `y`, i.e., `dim(r) = c(nrow(x)*nrow, ncol(x) + ncol(y))`.

If `all.x` is true, all the non matching cases of `x` are appended to the result as well, with `NA` filled in the corresponding columns of `y`; analogously for `all.y`.

If the remaining columns in the data frames have any common names, these have `suffixes` (`".x"` and `".y"` by default) appended to make the names of the result unique.

Value

A data frame. The rows are by default lexicographically sorted on the common columns, but are otherwise in the order in which they occurred in `y`. The columns are the common columns followed by the remaining columns in `x` and then those in `y`. If the matching involved row names, an extra column `Row.names` is added at the left, and in all cases the result has no special row names.

See Also

`data.frame`, `by`, `cbind`

Examples

```
authors <- data.frame(
  surname = c("Tukey", "Venables", "Tierney", "Ripley",
              "McNeil"),
  nationality = c("US", "Australia", "US", "UK",
                  "Australia"),
  deceased = c("yes", rep("no", 4)))
oks <- data.frame(
  name = c("Tukey", "Venables", "Tierney",
           "Ripley", "Ripley", "McNeil", "R Core"),
  title = c("Exploratory Data Analysis",
```

```
            "Modern Applied Statistics ...",
            "LISP-STAT",
            "Spatial Statistics", "Stochastic Simulation",
            "Interactive Data Analysis",
            "An Introduction to R"),
  other.author = c(NA, "Ripley", NA, NA, NA, NA,
                   "Venables & Smith"))

(m1 <- merge(authors, books,
             by.x = "surname", by.y = "name"))
(m2 <- merge(books, authors,
             by.x = "name", by.y = "surname"))
stopifnot(as.character(m1[,1]) == as.character(m2[,1]),
          all.equal(m1[, -1], m2[, -1][ names(m1)[-1] ]),
          dim(merge(m1, m2, by = integer(0))) == c(36, 10))

## "R core" is missing from authors and appears only here :
merge(authors, books,
      by.x = "surname", by.y = "name", all = TRUE)
```

`methods` *List Methods for S3 Generic Functions or Classes*

Description

List all available methods for an S3 generic function, or all methods for a class.

Usage

```
methods(generic.function, class)
```

Arguments

`generic.function`
: a generic function, or a character string naming a generic function.

`class`
: a symbol or character string naming a class: only used if `generic.function` is not supplied.

Details

Function `methods` can be used to find out about the methods for a particular generic function or class. The functions listed are those which *are named like methods* and may not actually be methods (known exceptions are discarded in the code). Note that the listed methods may not be user-visible objects, but often help will be available for them.

If `class` is used, we check that a matching generic can be found for each user-visible object named.

Value

An object of class `"MethodsFunction"`, a character vector of function names with an `"info"` attribute. There is a `print` method which marks with an asterisk any methods which are not visible: such functions can be examined by `getS3method` or `getAnywhere`.

The `"info"` attribute is a data frame, currently with a logical column, `visible` and a factor column `from` (indicating where the methods were found).

Note

This scheme is called *S3* (S version 3). For new projects, it is recommended to use the more flexible and robust *S4* scheme provided in the **methods** package. Functions can have both S3 and S4 methods, and function `showMethods` will list the S4 methods (possibly none).

The original `methods` function was written by Martin Maechler.

References

Chambers, J. M. (1992) *Classes and methods: object-oriented programming in S.* Appendix A of *Statistical Models in S* eds J. M. Chambers and T. J. Hastie, Wadsworth & Brooks/Cole.

See Also

`S3Methods`, `class`, `getS3method`

Examples

```
methods(summary)
methods(class = "aov")
methods("[[")    # does not list the C-internal ones...
methods("$")     # currently none
methods("$<-")   # replacement function
methods("+")     # binary operator
methods("Math")  # group generic
methods(print)
```

missing *Does a Formal Argument have a Value?*

Description

`missing` can be used to test whether a value was specified as an argument to a function.

Usage

```
missing(x)
```

Arguments

`x`	a formal argument.

Details

`missing(x)` is only reliable if `x` has not been altered since entering the function: in particular it will *always* be false after `x <- match.arg(x)`.

The example shows how a plotting function can be written to work with either a pair of vectors giving x and y coordinates of points to be plotted or a single vector giving y values to be plotted against their indexes.

Currently `missing` can only be used in the immediate body of the function that defines the argument, not in the body of a nested function or a `local` call. This may change in the future.

References

Becker, R. A., Chambers, J. M. and Wilks, A. R. (1988) *The New S Language.* Wadsworth & Brooks/Cole.

Chambers, J. M. (1998) *Programming with Data. A Guide to the S Language.* Springer.

See Also

`substitute` for argument expression; `NA` for "missing values" in data.

Examples

```
myplot <- function(x,y) {
                if(missing(y)) {
                        y <- x
                        x <- 1:length(y)
                }
                plot(x,y)
        }
```

mode *The (Storage) Mode of an Object*

Description

Get or set the type or storage mode of an object.

Usage

```
mode(x)
mode(x) <- value
storage.mode(x)
storage.mode(x) <- value
```

Arguments

`x`	any R object.
`value`	a character string giving the desired (storage) mode of the object.

Details

Both `mode` and `storage.mode` return a character string giving the (storage) mode of the object — often the same — both relying on the output of `typeof(x)`, see the example below.

The two assignment versions are currently identical. Both `mode(x) <- newmode` and `storage.mode(x) <- newmode` change the `mode` or `storage.mode` of object `x` to `newmode`.

As storage mode `"single"` is only a pseudo-mode in R, it will not be reported by `mode` or `storage.mode`: use `attr(object, "Csingle")` to examine this. However, the assignment versions can be used to set the mode to `"single"`, which sets the real mode to `"double"` and the `"Csingle"` attribute to `TRUE`. Setting any other mode will remove this attribute.

Note (in the examples below) that some `calls` have mode `"("` which is S compatible.

References

Becker, R. A., Chambers, J. M. and Wilks, A. R. (1988) *The New S Language*. Wadsworth & Brooks/Cole.

See Also

`typeof` for the R-internal "mode", `attributes`.

Examples

```
sapply(options(),mode)

cex3 <- c("NULL","1","1:1","1i","list(1)",
  "data.frame(x=1)", "pairlist(pi)",
  "c", "lm", "formals(lm)[[1]]", "formals(lm)[[2]]",
  "y~x","expression((1))[[1]]", "(y~x)[[1]]",
  "expression(x <- pi)[[1]][[1]]")
lex3 <- sapply(cex3, function(x) eval(parse(text=x)))
mex3 <- t(sapply(lex3, function(x)
          c(typeof(x), storage.mode(x), mode(x))))
dimnames(mex3) <- list(cex3,
          c("typeof(.)","storage.mode(.)","mode(.)"))
mex3

## This also makes a local copy of 'pi':
storage.mode(pi) <- "complex"
storage.mode(pi)
rm(pi)
```

NA *Not Available / "Missing" Values*

Description

`NA` is a logical constant of length 1 which contains a missing value indicator. `NA` can be freely coerced to any other vector type.

The generic function `is.na` indicates which elements are missing.

The generic function `is.na<-` sets elements to `NA`.

Usage

```
NA
is.na(x)
## S3 method for class 'data.frame':
is.na(x)

is.na(x) <- value
```

Arguments

`x`	an R object to be tested.
`value`	a suitable index vector for use with `x`.

Details

The `NA` of character type is as from R 1.5.0 distinct from the string `"NA"`. Programmers who need to specify an explicit string `NA` should use `as.character(NA)` rather than `"NA"`, or set elements to `NA` using `is.na<-`.

`is.na(x)` works elementwise when `x` is a `list`. The method dispatching is C-internal, rather than via `UseMethod`.

Function `is.na<-` may provide a safer way to set missingness. It behaves differently for factors, for example.

Value

The default method for `is.na` returns a logical vector of the same "form" as its argument `x`, containing `TRUE` for those elements marked `NA` or `NaN` (!) and `FALSE` otherwise. `dim`, `dimnames` and `names` attributes are preserved.

The method `is.na.data.frame` returns a logical matrix with the same dimensions as the data frame, and with dimnames taken from the row and column names of the data frame.

References

Becker, R. A., Chambers, J. M. and Wilks, A. R. (1988) *The New S Language.* Wadsworth & Brooks/Cole.

Chambers, J. M. (1998) *Programming with Data. A Guide to the S Language.* Springer.

See Also

`NaN`, `is.nan`, etc., and the utility function `complete.cases`.

`na.action`, `na.omit`, `na.fail` on how methods can be tuned to deal with missing values.

Examples

```
is.na(c(1, NA))        # FALSE  TRUE
is.na(paste(c(1, NA))) # FALSE FALSE
```

`na.action` *NA Action*

Description

`na.action` is a generic function, and `na.action.default` its default method.

Usage

```
na.action(object, ...)
```

Arguments

`object`	any object whose `NA` action is given.
`...`	further arguments special methods could require.

Value

The "NA action" which should be applied to `object` whenever `NA`s are not desired.

References

Chambers, J. M. and Hastie, T. J. (1992) *Statistical Models in S.* Wadsworth & Brooks/Cole.

See Also

`options("na.action")`, `na.omit`, `na.fail`

Examples

```
na.action(c(1, NA))
```

na.fail *Handle Missing Values in Objects*

Description

These generic functions are useful for dealing with NAs in e.g., data frames. na.fail returns the object if it does not contain any missing values, and signals an error otherwise. na.omit returns the object with incomplete cases removed. na.pass returns the object unchanged.

Usage

```
na.fail(object, ...)
na.omit(object, ...)
na.exclude(object, ...)
na.pass(object, ...)
```

Arguments

object	an R object, typically a data frame
...	further arguments special methods could require.

Details

At present these will handle vectors, matrices and data frames comprising vectors and matrices (only).

If na.omit removes cases, the row numbers of the cases form the "na.action" attribute of the result, of class "omit".

na.exclude differs from na.omit only in the class of the "na.action" attribute of the result, which is "exclude". This gives different behaviour in functions making use of naresid and napredict: when na.exclude is used the residuals and predictions are padded to the correct length by inserting NAs for cases omitted by na.exclude.

References

Chambers, J. M. and Hastie, T. J. (1992) *Statistical Models in S.* Wadsworth & Brooks/Cole.

See Also

na.action; options with argument na.action for setting “NA actions”; and lm and glm for functions using these.

Examples

```
DF <- data.frame(x = c(1, 2, 3), y = c(0, 10, NA))
na.omit(DF)
m <- as.matrix(DF)
na.omit(m)
# does not affect objects with no NA's
stopifnot(all(na.omit(1:3) == 1:3))
try(na.fail(DF)) # Error: missing values in ...

options("na.action")
```

name *Variable Names or Symbols, respectively*

Description

`as.symbol` coerces its argument to be a *symbol*, or equivalently, a *name*. The argument must be of mode `"character"`. `as.name` is an alias for `as.symbol`.

`is.symbol` (and `is.name` equivalently) returns `TRUE` or `FALSE` depending on whether its argument is a symbol (i.e., name) or not.

Usage

```
as.symbol(x)
is.symbol(y)

as.name(x)
is.name(y)
```

Arguments

`x, y`	objects to be coerced or tested.

Details

`is.symbol` is generic: you can write methods to handle specific classes of objects, see InternalMethods.

Note

The term "symbol" is from the LISP background of R, whereas "name" has been the standard S term for this.

References

Becker, R. A., Chambers, J. M. and Wilks, A. R. (1988) *The New S Language*. Wadsworth & Brooks/Cole.

See Also

`call`, `is.language`. For the internal object mode, `typeof`.

Examples

```
an <- as.name("arrg")
is.name(an) # TRUE
str(an) # symbol
```

names *The Names Attribute of an Object*

Description

Functions to get or set the names of an object.

Usage

```
names(x)
names(x) <- value
```

Arguments

`x`	an R object.
`value`	a character vector of up to the same length as `x`, or `NULL`.

Details

`names` is a generic accessor function, and `names<-` is a generic replacement function. The default methods get and set the `"names"` attribute of a vector or list.

If `value` is shorter than `x`, it is extended by character `NA`s to the length of `x`.

It is possible to update just part of the names attribute via the general rules: see the examples. This works because the expression there is evaluated as `z <- "names<-"(z, "[<-"(names(z), 3, "c2"))`.

Value

For `names`, `NULL` or a character vector of the same length as `x`.

For `names<-`, the updated object. (Note that the value of `names(x) <- value` is that of the assignment, `value`, not the return value from the left-hand side.)

References

Becker, R. A., Chambers, J. M. and Wilks, A. R. (1988) *The New S Language.* Wadsworth & Brooks/Cole.

Examples

```
data(islands)
# print the names attribute of the islands data set
names(islands)

# remove the names attribute
names(islands) <- NULL

z <- list(a=1, b="c", c=1:3)
names(z)
# change just the name of the third element.
names(z)[3] <- "c2"
z

## assign just one name
z <- 1:3
names(z)
# change just the name of the third element.
names(z)[2] <- "b"
z
```

nargs *The Number of Arguments to a Function*

Description

When used inside a function body, **nargs** returns the number of arguments supplied to that function, *including* positional arguments left blank.

Usage

```
nargs()
```

References

Becker, R. A., Chambers, J. M. and Wilks, A. R. (1988) *The New S Language*. Wadsworth & Brooks/Cole.

See Also

`args`, `formals` and `sys.call`.

Examples

```
tst <- function(a, b = 3, ...) {nargs()}
tst() # 0
tst(clicketyclack) # 1 (even non-existing)
tst(c1, a2, rr3) # 3

foo <- function(x, y, z, w) {
   cat("call was", deparse(match.call()), "\n")
   nargs()
}
foo()    # 0
foo(,,3) # 3
foo(z=3) # 1, even though this is the same call

nargs() # not really meaningful
```

nchar *Count the Number of Characters*

Description

`nchar` takes a character vector as an argument and returns a vector whose elements contain the number of characters in the corresponding element of `x`.

Usage

```
nchar(x)
```

Arguments

`x`	character vector, or a vector to be coerced to a character vector.

Details

The internal equivalent of `as.character` is performed on `x`. If you want to operate on non-vector objects passing them through `deparse` first will be required.

Value

The number of characters as the string will be printed (integer 2 for a missing string).

References

Becker, R. A., Chambers, J. M. and Wilks, A. R. (1988) *The New S Language*. Wadsworth & Brooks/Cole.

See Also

`strwidth` giving width of strings for plotting; `paste`, `substr`, `strsplit`

Examples

```
x <- c("asfef","qwerty","yuiop[","b","stuff.blah.yech")
nchar(x)
# 5  6  6  1 15
```

```
nchar(deparse(mean))
# 23  1 16 45 11 64  2 17 50 43  2 17  1
```

nclass *Compute the Number of Classes for a Histogram*

Description

Compute the number of classes for a histogram, for use internally in `hist`.

Usage

```
nclass.Sturges(x)
nclass.scott(x)
nclass.FD(x)
```

Arguments

`x`	A data vector.

Details

`nclass.Sturges` uses Sturges' formula, implicitly basing bin sizes on the range of the data.

`nclass.scott` uses Scott's choice for a normal distribution based on the estimate of the standard error.

`nclass.FD` uses the Freedman-Diaconis choice based on the inter-quartile range.

Value

The suggested number of classes.

References

Venables, W. N. and Ripley, B. D. (2002) *Modern Applied Statistics with S-PLUS.* Springer, page 112.

Freedman, D. and Diaconis, P. (1981) On the histogram as a density estimator: L_2 theory. *Zeitschrift für Wahrscheinlichkeitstheorie und verwandte Gebiete* **57**, 453–476.

Scott, D. W. (1979) On optimal and data-based histograms. *Biometrika* **66**, 605–610.

Scott, D. W. (1992) *Multivariate Density Estimation. Theory, Practice, and Visualization.* Wiley.

See Also

`hist`

`nlevels` *The Number of Levels of a Factor*

Description

Return the number of levels which its argument has.

Usage

```
nlevels(x)
```

Arguments

`x`	an object, usually a factor.

Details

If the argument is not a `factor`, `NA` is returned.

The actual factor levels (if they exist) can be obtained with the `levels` function.

Examples

```
nlevels(gl(3,7)) # = 3
```

noquote *Class for "no quote" Printing of Character Strings*

Description

Print character strings without quotes.

Usage

```
noquote(obj)
## S3 method for class 'noquote':
print(x, ...)
## S3 method for class 'noquote':
c(..., recursive = FALSE)
```

Arguments

`obj`	any R object, typically a vector of `character` strings.
`x`	an object of class `"noquote"`.
`...`	further options passed to next methods, such as `print`.
`recursive`	for compatibility with the generic `c` function.

Details

`noquote` returns its argument as an object of class `"noquote"`. There is a method for `c()` and subscript method (`"[.noquote"`) which ensures that the class is not lost by subsetting. The print method (`print.noquote`) prints character strings *without* quotes (`"..."`).

These functions exist both as utilities and as an example of using (S3) `class` and object orientation.

Author(s)

Martin Maechler

See Also

`methods`, `class`, `print`.

Examples

```
letters
nql <- noquote(letters)
nql
nql[1:4] <- "oh"
nql[1:12]

cmp.logical <- function(log.v)
{
  ## Purpose: compact printing of logicals
  log.v <- as.logical(log.v)
  noquote(if(length(log.v)==0)"()" else c(".","|")[1+log.v])
}
cmp.logical(runif(20) > 0.8)
```

NotYet *Not Yet Implemented Functions and Unused Arguments*

Description

In order to pinpoint missing functionality, the R core team uses these functions for missing R functions and not yet used arguments of existing R functions (which are typically there for compatibility purposes).

You are very welcome to contribute your code ...

Usage

```
.NotYetImplemented()
.NotYetUsed(arg, error = TRUE)
```

Arguments

`arg`	an argument of a function that is not yet used.
`error`	a logical. If `TRUE`, an error is signalled; if `FALSE`; only a warning is given.

See Also

the contrary, `Deprecated` and `Defunct` for outdated code.

Examples

```
plot.mlm        # to see how the "NotYetImplemented"
                # reference is made automagically
try(plot.mlm())

barplot(1:5, inside = TRUE) # 'inside' is not yet used
```

nrow *The Number of Rows/Columns of an Array*

Description

`nrow` and `ncol` return the number of rows or columns present in `x`. `NCOL` and `NROW` do the same treating a vector as 1-column matrix.

Usage

```
nrow(x)
ncol(x)
NCOL(x)
NROW(x)
```

Arguments

`x`	a vector, array or data frame

Value

an `integer` of length 1 or `NULL`.

References

Becker, R. A., Chambers, J. M. and Wilks, A. R. (1988) *The New S Language*. Wadsworth & Brooks/Cole (`ncol` and `nrow`.)

See Also

`dim` which returns *all* dimensions; `array`, `matrix`.

Examples

```
ma <- matrix(1:12, 3, 4)
nrow(ma)   # 3
ncol(ma)   # 4

ncol(array(1:24, dim = 2:4)) # 3, the second dimension
NCOL(1:12) # 1
NROW(1:12) # 12
```

`ns-alt` *Experimental Alternative Name Specification Support*

Description

Alternative interface for specifying a name space within the code of a package.

Usage

```
.Export(...)
.Import(...)
.ImportFrom(name, ...)
.S3method(generic, class, method)
```

Arguments

`...`	name or literal character string arguments.
`name`	name or literal character string.
`generic`	name or literal character string.
`class`	name or literal character string.
`method`	optional character or function argument.

Details

As an experimental alternative to using a '`NAMESPACE`' file it is possible to add a name space to a package by adding a `Namespace: <package_name>` entry to the '`DESCRIPTION`' file and placing directives to specify imports and exports directly in package code. These directives should be viewed as declarations, not as function calls. Except to the optional method argument to `.S3method` arguments are not evaluated. These directives should only be used at top level of package code except as noted below.

`.Export` is used to declare exports. Its arguments should be literal names or character strings. `.Export` should only be used at package top level.

`.Import` is used to declare the import of entire name spaces. Its arguments should be literal names or character strings. `.ImportFrom` is used to declare the import of selected variables from a single name space. The first argument is a literal name or character string identifying the source name space; the remaining arguments are literal names or character strings identifying the variables to import. As an experimental feature both `.Import` and `.ImportFrom` can be used to import variables into a local environment. The drawback of allowing this is that dependencies cannot be determined easily at package load time, and as a result this feature may need to be dropped.

`.S3method` is used to declare a method for S3-style `UseMethod` dispatch. This is needed since methods in packages that are imported but not on the search path might not be visible to the standard dispatch mechanism at a call site. The first argument is the name of the generic, the second specifies the class. The third argument is optional and defaults to the usual concatenation of generic and class separated by a period. If supplied, the third argument should evaluate to a character string or a function. If the third argument is omitted or a character string is supplied, then a function by that name must be defined. If a function is supplied, it is used as the method. When the method is specified as a name, explicitly or implicitly, the function lookup is handled lazily; this allows the definition to occur after the `.S3method` declaration and also integrates with possible database storage of package code.

Author(s)

Luke Tierney

Examples

```
## code for package/name space 'foo'
x <- 1
f <- function(y) c(x,y)
print.foo <- function(x, ...) cat("<a foo>\n")
.Export(f)
S3method(print,foo)

## code for package/name space 'bar'
.Import(foo)
c <- function(...) sum(...)
g <- function(y) f(c(y, 7))
h <- function(y) y+9
.Export(g, h)
```

`ns-dblcolon` *Double Colon and Triple Colon Operators*

Description

Accessing exported and internal variables in a name space.

Usage

```
pkg::name
pkg:::name
```

Arguments

`pkg`	package name symbol or literal character string.
`name`	variable name symbol or literal character string.

Details

The expression `pkg::name` returns the value of the exported variable `name` in package `pkg` if the package has a name space. The expression `pkg:::name` returns the value of the internal variable `name` in package `pkg` if the package has a name space. The package will be loaded if it was not loaded already before the call. Assignment into name spaces is not supported.

Examples

```
base::log
base::"+"
```

ns-internals *Name Space Internals*

Description

Internal name space support functions. Not intended to be called directly.

Usage

```
asNamespace(ns, base.OK = TRUE)
getNamespaceInfo(ns, which)
importIntoEnv(impenv, impnames, expenv, expnames)
isBaseNamespace(ns)
namespaceExport(ns, vars)
namespaceImport(self, ...)
namespaceImportFrom(self, ns, vars)
namespaceImportClasses(self, ns, vars)
namespaceImportMethods(self, ns, vars)
packageHasNamespace(package, package.lib)
parseNamespaceFile(package, package.lib, mustExist = TRUE)
registerS3method(genname, class, method,
                 envir = parent.frame())
setNamespaceInfo(ns, which, val)
.mergeExportMethods(new, ns)
```

Arguments

`ns`	string or name space environment.
`base.OK`	logical.
`impenv`	environment.
`expenv`	name space environment.
`vars`	character vector.
`self`	name space environment.
`package`	string naming the package/name space to load.
`package.lib`	character vector specifying library.
`mustExist`	logical.
`genname`	character.
`class`	character.

`envir`	environment.
`which`	character.
`val`	any object.
`...`	character arguments.

Author(s)

Luke Tierney

ns-lowlev *Low Level Name Space Support Functions*

Description

Low level name space support functions.

Usage

```
attachNamespace(ns, pos = 2)
loadNamespace(package, lib.loc = NULL,
              keep.source = getOption("keep.source.pkgs"),
              partial = FALSE, declarativeOnly = FALSE)
loadedNamespaces()
unloadNamespace(ns)
loadingNamespaceInfo()
saveNamespaceImage(package, rdafile, lib.loc = NULL,
  keep.source = getOption("keep.source.pkgs"))
```

Arguments

`ns`	string or namespace object.
`pos`	integer specifying position to attach.
`package`	string naming the package/name space to load.
`lib.loc`	character vector specifying library search path.
`keep.source`	logical specifying whether to retain source.
`partial`	logical; if true, stop just after loading code.
`declarativeOnly`	logical; disables `.Import`, etc, if true.

Details

The functions `loadNamespace` and `attachNamespace` are usually called implicitly when `library` is used to load a name space and any imports needed. However it may be useful to call these functions directly at times.

`loadNamespace` loads the specified name space and registers it in an internal database. A request to load a name space that is already loaded has no effect. The arguments have the same meaning as the corresponding arguments to `library`. After loading, `loadNamespace` looks for a hook function named `.onLoad` as an internal variable in the name space (it should not be exported). This function is called with the same arguments as `.First.lib`. Partial loading is used so support installation with the '`--save`' option.

`loadNamespace` does not attach the name space it loads to the search path. `attachNamespace` can be used to attach a frame containing the exported values of a name space to the search path. The hook function `.onAttach` is run after the name space exports are attached, but this is not likely to be useful. Shared library loading and setting of options should be handled at load time by the `.onLoad` hook.

`loadedNamespaces` returns a character vector of the names of the loaded name spaces.

`unloadNamespace` can be used to force a name space to be unloaded. An error is signaled if the name space is imported by other loaded name spaces. If defined, a hook function `.onUnload`, analogous to `.Last.lib`, is run before removing the name space from the internal registry. `unloadNamespace` will first `detach` a package of the same name if one is on the path, thereby running a `.Last.lib` function in the package if one is exported.

`loadingNamespaceInfo` returns a list of the arguments that would be passed to `.onLoad` when a name space is being loaded. An error is signaled of a name space is not currently being loaded.

`saveNamespaceImage` is used to save name space images for packages installed with '`--save`'.

Author(s)

Luke Tierney

ns-reflect.Rd *Name Space Reflection Support*

Description

Functions to support reflection on name space objects.

Usage

```
getExportedValue(ns, name)
getNamespace(name)
getNamespaceExports(ns)
getNamespaceImports(ns)
getNamespaceName(ns)
getNamespaceUsers(ns)
getNamespaceVersion(ns)
```

Arguments

`ns`	string or name space object.
`name`	string or name.

Details

`getExportedValue` returns the value of the exported variable `name` in name space `ns`.

`getNamespace` returns the environment representing the name space `name`. The name space is loaded if necessary.

`getNamespaceExports` returns a character vector of the names exported by `ns`.

`getNamespaceImports` returns a representation of the imports used by name space `ns`. This representation is experimental and subject to change.

`getNamespaceName` and `getNamespaceVersion` return the name and version of the name space `ns`.

`getNamespaceUsers` returns a character vector of the names of the name spaces that import name space `ns`.

Author(s)

Luke Tierney

`ns-topenv` *Top Level Environment*

Description

Finding the top level environment.

Usage

```
topenv(envir = parent.frame(),
       matchThisEnv = getOption("topLevelEnvironment"))
```

Arguments

`envir`	environment.
`matchThisEnv`	return this environment, if it matches before any other criterion is satisfied. The default, the option "topLevelEnvironment", is set by `sys.source`, which treats a specific environment as the top level environment. Supplying the argument as `NULL` means it will never match.

Details

`topenv` returns the first top level environment found when searching `envir` and its parent environments. An environment is considered top level if it is the internal environment of a name space, a package environment in the search path, or `.GlobalEnv`.

Examples

```
topenv(.GlobalEnv)
topenv(new.env())
```

`nsl` *Look up the IP Address by Hostname*

Description

Interface to `gethostbyname`.

Usage

```
nsl(hostname)
```

Arguments

`hostname`	the name of the host.

Value

The IP address, as a character string, or `NULL` if the call fails.

Note

This was included as a test of internet connectivity, to fail if the node running R is not connected. It will also return `NULL` if BSD networking is not supported, including the header file '`arpa/inet.h`'.

Examples

```
nsl("www.r-project.org")
```

NULL *The Null Object*

Description

`NULL` represents the null object in R. NULL is used mainly to represent the lists with zero length, and is often returned by expressions and functions whose value is undefined.

`as.null` ignores its argument and returns the value `NULL`.

`is.null` returns `TRUE` if its argument is `NULL` and `FALSE` otherwise.

Usage

```
NULL
as.null(x, ...)
is.null(x)
```

Arguments

`x`	an object to be tested or coerced.
`...`	ignored.

Details

`is.null` is generic: you can write methods to handle specific classes of objects, see InternalMethods.

References

Becker, R. A., Chambers, J. M. and Wilks, A. R. (1988) *The New S Language.* Wadsworth & Brooks/Cole.

Examples

```
is.null(list())    # FALSE (on purpose!)
is.null(integer(0)) # F
is.null(logical(0)) # F
as.null(list(a=1,b='c'))
```

`numeric` *Numeric Vectors*

Description

`numeric` creates a real vector of the specified length. The elements of the vector are all equal to `0`.

`as.numeric` attempts to coerce its argument to numeric type (either integer or real).

`is.numeric` returns `TRUE` if its argument is of type real or type integer and `FALSE` otherwise.

Usage

```
numeric(length = 0)
as.numeric(x, ...)
is.numeric(x)
```

Arguments

`length`	desired length.
`x`	object to be coerced or tested.
`...`	further arguments passed to or from other methods.

Details

`is.numeric` is generic: you can write methods to handle specific classes of objects, see InternalMethods.

Note that factors are false for `is.numeric` but true for `is.integer`.

Note

R *has no single precision data type. All real numbers are stored in double precision format.* While `as.numeric` is a generic function, user methods must be written for `as.double`, which it calls

`as.numeric` for factors yields the codes underlying the factor levels, not the numeric representation of the labels.

References

Becker, R. A., Chambers, J. M. and Wilks, A. R. (1988) *The New S Language.* Wadsworth & Brooks/Cole.

Examples

```
# (-0.1, 2.7, NA)  +  warning
as.numeric(c("-.1"," 2.7 ","B"))
as.numeric(factor(5:10))
```

object.size *Report the Space Allocated for an Object*

Description

Provides an estimate of the memory that is being used to store an R object.

Usage

```
object.size(x)
```

Arguments

`x`	An R object.

Details

Exactly which parts of the memory allocation should be attributed to which object is not clear-cut. This function merely provides a rough indication. For example, it will not detect if character storage for character strings are shared between identical elements (which it will be if `rep` was used, for example).

The calculation is of the size of the object, and excludes the space needed to store its name in the symbol table.

Value

An estimate of the memory allocation attributable to the object, in bytes.

Examples

```
object.size(letters)
object.size(ls)
## find the 10 largest objects in base
z <- sapply(ls("package:base"),
            function(x) object.size(get(x, envir=NULL)))
as.matrix(rev(sort(z))[1:10])
```

`octmode` *Display Numbers in Octal*

Description

Convert or print integers in octal format, with as many digits as are needed to display the largest, using leading zeroes as necessary.

Usage

```
## S3 method for class 'octmode':
as.character(x, ...)
## S3 method for class 'octmode':
format(x, ...)
## S3 method for class 'octmode':
print(x, ...)
```

Arguments

`x`	An object inheriting from class `"octmode"`.
`...`	further arguments passed to or from other methods.

Details

Class `"octmode"` consists of integer vectors with that class attribute, used merely to ensure that they are printed in octal notation, specifically for Unix-like file permissions such as `755`.

See Also

These are auxiliary functions for `file.info`

`on.exit` *Function Exit Code*

Description

`on.exit` records the expression given as its argument as needing to be executed when the current function exits (either naturally or as the result of an error). This is useful for resetting graphical parameters or performing other cleanup actions.

If no expression is provided, i.e., the call is `on.exit()`, then the current `on.exit` code is removed.

Usage

```
on.exit(expr, add = FALSE)
```

Arguments

`expr`	an expression to be executed.
`add`	if TRUE, add `expr` to be executed after any previously set expressions.

References

Becker, R. A., Chambers, J. M. and Wilks, A. R. (1988) *The New S Language.* Wadsworth & Brooks/Cole.

See Also

`sys.on.exit` to see the current expression.

Examples

```
opar <- par(mai = c(1,1,1,1))
on.exit(par(opar))
```

options *Options Settings*

Description

Allow the user to set and examine a variety of global "options" which affect the way in which R computes and displays its results.

Usage

```
options(...)
getOption(x)
.Options
```

Arguments

`...`	any options can be defined, using `name = value`. However, only the ones below are used in "base R". Further, `options('name') == options()['name']`, see the example.
`x`	a character string holding an option name.

Details

Invoking `options()` with no arguments returns a list with the current values of the options. Note that not all options listed below are set initially. To access the value of a single option, one should use `getOption("width")`, e.g., rather than `options("width")` which is a *list* of length one.

`.Options` also always contains the `options()` list, for S compatibility. You must use it "read only" however.

Value

For `options`, a list (in any case) with the previous values of the options changed, or all options when no arguments were given.

Options used in base R

`prompt`: a string, used for R's prompt; should usually end in a blank (`" "`).

continue: a string setting the prompt used for lines which continue over one line.

width: controls the number of characters on a line. You may want to change this if you re-size the window that R is running in. Valid values are 10...10000 with default normally 80. (The valid values are in file '`Print.h`' and can be changed by re-compiling R.)

digits: controls the number of digits to print when printing numeric values. It is a suggestion only. Valid values are 1...22 with default 7. See `print.default`.

editor: sets the default text editor, e.g., for `edit`. Set from the environment variable `VISUAL` on UNIX.

pager: the (stand-alone) program used for displaying ASCII files on R's console, also used by `file.show` and sometimes `help`. Defaults to '`$R_HOME/bin/pager`'.

browser: default HTML browser used by `help.start()` on UNIX, or a non-default browser on Windows.

pdfviewer: default PDF viewer. Set from the environment variable `R_PDFVIEWER`.

mailer: default mailer used by `bug.report()`. Can be `"none"`.

contrasts: the default `contrasts` used in model fitting such as with `aov` or `lm`. A character vector of length two, the first giving the function to be used with unordered factors and the second the function to be used with ordered factors.

defaultPackages: the packages that are attached by default when R starts up. Initially set from value of the environment variables `R_DefaultPackages`, or if that is unset to `c("ts", "nls", "modreg", "mva", "ctest", "methods")`. (Set `R_DEFAULT_PACKAGES` to `NULL` or a comma-separated list of package names.) A call to `options` should be in your '`.Rprofile`' file to ensure that the change takes effect before the base package is initialized (see `Startup`).

expressions: sets a limit on the number of nested expressions that will be evaluated. Valid values are 25...100000 with default 500.

keep.source: When `TRUE`, the source code for functions (newly defined or loaded) is stored in their `"source"` attribute (see `attr`) allowing comments to be kept in the right places.

The default is `interactive()`, i.e., `TRUE` for interactive use.

keep.source.pkgs: As for `keep.source`, for functions in packages loaded by `library` or `require`. Defaults to `FALSE` unless the environment variable `R_KEEP_PKG_SOURCE` is set to `yes`.

na.action: the name of a function for treating missing values (NA's) for certain situations.

papersize: the default paper format used by postscript; set by environment variable R_PAPERSIZE when R is started and defaulting to "a4" if that is unset or invalid.

printcmd: the command used by postscript for printing; set by environment variable R_PRINTCMD when R is started. This should be a command that expects either input to be piped to 'stdin' or to be given a single filename argument.

latexcmd, dvipscmd: character strings giving commands to be used in off-line printing of help pages.

show.signif.stars, show.coef.Pvalues: logical, affecting P value printing, see print.coefmat.

ts.eps: the relative tolerance for certain time series (ts) computations.

error: either a function or an expression governing the handling of non-catastrophic errors such as those generated by stop as well as by signals and internally detected errors. If the option is a function, a call to that function, with no arguments, is generated as the expression. The default value is NULL: see stop for the behaviour in that case. The function dump.frames provides one alternative that allows post-mortem debugging.

show.error.messages: a logical. Should error messages be printed? Intended for use with try or a user-installed error handler.

warn: sets the handling of warning messages. If warn is negative all warnings are ignored. If warn is zero (the default) warnings are stored until the top–level function returns. If fewer than 10 warnings were signalled they will be printed otherwise a message saying how many (max 50) were signalled. A top–level variable called last.warning is created and can be viewed through the function warnings. If warn is one, warnings are printed as they occur. If warn is two or larger all warnings are turned into errors.

warning.length: sets the truncation limit for error and warning messages. A non-negative integer, with allowed values 100–8192, default 1000.

warning.expression: an R code expression to be called if a warning is generated, replacing the standard message. If non-null is called irrespective of the value of option warn.

check.bounds: logical, defaulting to FALSE. If true, a warning is produced whenever a "generalized vector" (atomic or list) is extended, by something like x <- 1:3; x[5] <- 6.

echo: logical. Only used in non-interactive mode, when it controls whether input is echoed. Command-line option '--slave' sets this initially to FALSE.

verbose: logical. Should R report extra information on progress? Set to TRUE by the command-line option '--verbose'.

device: a character string giving the default device for that session. This defaults to the normal screen device (e.g., x11, windows or gtk) for an interactive session, and postscript in batch use or if a screen is not available.

X11colortype: The default colour type for X11 devices.

CRAN: The URL of the preferred CRAN node for use by update.packages. Defaults to http://cran.r-project.org.

download.file.method: Method to be used for download.file. Currently download methods "internal", "wget" and "lynx" are available. There is no default for this option, when method = "auto" is chosen: see download.file.

unzip: the command used for unzipping help files. Defaults to the value of R_UNZIPCMD, which is set in 'etc/Renviron' if an unzip command was found during configuration.

de.cellwidth: integer: the cell widths (number of characters) to be used in the data editor dataentry. If this is unset, 0, negative or NA, variable cell widths are used.

encoding: An integer vector of length 256 holding an input encoding. Defaults to native.enc (= 0:255). See connections.

timeout: integer. The timeout for some Internet operations, in seconds. Default 60 seconds. See download.file and connections.

internet.info: The minimum level of information to be printed on URL downloads etc. Default is 2, for failure causes. Set to 1 or 0 to get more information.

scipen: integer. A penalty to be applied when deciding to print numeric values in fixed or exponential notation. Positive values bias towards fixed and negative towards scientific notation: fixed notation will be preferred unless it is more than scipen digits wider.

locatorBell: logical. Should selection in locator and identify be confirmed by a bell. Default TRUE. Honoured at least on X11 and windows devices.

The default settings of some of these options are

prompt	"> "	continue	"+ "

width	80	digits	7
expressions	500	keep.source	TRUE
show.signif.stars	TRUE	show.coef.Pvalues	TRUE
na.action	na.omit	ts.eps	1e-5
error	NULL	show.error.messages	TRUE
warn	0	warning.length	1000
echo	TRUE	verbose	FALSE
scipen	0	locatorBell	TRUE

Others are set from environment variables or are platform-dependent.

References

Becker, R. A., Chambers, J. M. and Wilks, A. R. (1988) *The New S Language.* Wadsworth & Brooks/Cole.

Examples

```
options() # printing all current options
op <- options(); str(op) # nicer printing

# .Options is the same:
all(sapply(1:length(op),
      function(i) all(.Options[[i]] == op[[i]])))

# the latter needs more memory
options('width')[[1]] == options()$width
options(digits=20)
pi

# set the editor, and save previous value
old.o <- options(editor="nedit")
old.o

options(check.bounds = TRUE)
x <- NULL; x[4] <- "yes" # gives a warning

options(digits=5)
print(1e5)
options(scipen=3); print(1e5)

options(op)     # reset (all) initial options
options('digits')
```

```
## set contrast handling to be like S
options(contrasts=c("contr.helmert", "contr.poly"))

## on error, terminate the R session with error status 66
options(error=quote(q("no", status=66, runLast=FALSE)))
stop("test it")

## set an error action for debugging: see ?debugger.
options(error=dump.frames)
## A possible setting for non-interactive sessions
options(error=quote({dump.frames(to.file=TRUE); q()}))
```

order *Ordering Permutation*

Description

`order` returns a permutation which rearranges its first argument into ascending or descending order, breaking ties by further arguments. `sort.list` is the same, using only one argument.

Usage

```
order(..., na.last = TRUE, decreasing = FALSE)

sort.list(x, partial = NULL, na.last = TRUE,
          decreasing = FALSE,
          method = c("shell", "quick", "radix"))
```

Arguments

`...`	a sequence of vectors, all of the same length.
`x`	a vector.
`partial`	vector of indices for partial sorting.
`decreasing`	logical. Should the sort order be increasing or decreasing?
`na.last`	for controlling the treatment of `NA`s. If `TRUE`, missing values in the data are put last; if `FALSE`, they are put first; if `NA`, they are removed.
`method`	the method to be used: partial matches are allowed.

Details

In the case of ties in the first vector, values in the second are used to break the ties. If the values are still tied, values in the later arguments are used to break the tie (see the first example). The sort used is *stable* (except for `method = "quick"`), so any unresolved ties will be left in their original ordering.

The default method for `sort.list` is a good compromise. Method `"quick"` is only supported for numeric `x` with `na.last=NA`, and is not stable, but will be faster for long vectors. Method `"radix"` is only implemented for integer `x` with a range of less than 100,000. For such `x` it is very fast (and stable), and hence is ideal for sorting factors.

`partial` is supplied for compatibility with other implementations of S, but no other values are accepted and ordering is always complete.

References

Becker, R. A., Chambers, J. M. and Wilks, A. R. (1988) *The New S Language*. Wadsworth & Brooks/Cole.

See Also

`sort` and `rank`.

Examples

```
(ii <- order(x <- c(1,1,3:1,1:4,3), y <- c(9,9:1),
             z <-c(2,1:9)))
## 6  5  2  1  7  4 10 8  3  9
# shows the reordering (ties via 2nd & 3rd arg)
rbind(x,y,z)[,ii]

## Suppose we wanted descending order on y. A simple
## solution is
rbind(x,y,z)[, order(x, -y, z)]
## For character vectors we can make use of rank:
cy <- as.character(y)
rbind(x,y,z)[, order(x, -rank(y), z)]

## rearrange matched vectors so that the first is in
## ascending order
x <- c(5:1, 6:8, 12:9)
y <- (x - 5)^2
o <- order(x)
rbind(x[o], y[o])

## tests of na.last
a <- c(4, 3, 2, NA, 1)
b <- c(4, NA, 2, 7, 1)
z <- cbind(a, b)
(o <- order(a, b)); z[o, ]
(o <- order(a, b, na.last = FALSE)); z[o, ]
(o <- order(a, b, na.last = NA)); z[o, ]

## speed examples for long vectors: timings are
## immediately after gc()
```

```
x <- factor(sample(letters, 1e6, replace=TRUE))
system.time(o <- sort.list(x)) ## 4 secs
stopifnot(!is.unsorted(x[o]))
system.time(o <- sort.list(x, method="quick",
                           na.last=NA)) # 0.4 sec
stopifnot(!is.unsorted(x[o]))
system.time(o <- sort.list(x, method="radix")) # 0.04 sec
stopifnot(!is.unsorted(x[o]))
xx <- sample(1:26, 1e7, replace=TRUE)
system.time(o <- sort.list(xx, method="radix")) # 0.4 sec
xx <- sample(1:100000, 1e7, replace=TRUE)
system.time(o <- sort.list(xx, method="radix")) # 4 sec
```

outer *Outer Product of Arrays*

Description

The outer product of the arrays `X` and `Y` is the array `A` with dimension `c(dim(X), dim(Y))` where element `A[c(arrayindex.x, arrayindex.y)] = FUN(X[arrayindex.x], Y[arrayindex.y], ...)`.

Usage

```
outer(X, Y, FUN="*", ...)
X %o% Y
```

Arguments

`X`	A vector or array.
`Y`	A vector or array.
`FUN`	a function to use on the outer products, it may be a quoted string.
`...`	optional arguments to be passed to `FUN`.

Details

`FUN` must be a function (or the name of it) which expects at least two arguments and which operates elementwise on arrays.

Where they exist, the [dim]names of `X` and `Y` will be preserved.

`%o%` is an alias for `outer` (where `FUN` cannot be changed from `"*"`).

Author(s)

Jonathan Rougier

References

Becker, R. A., Chambers, J. M. and Wilks, A. R. (1988) *The New S Language.* Wadsworth & Brooks/Cole.

See Also

`%*%` for usual (*inner*) matrix vector multiplication; `kronecker` which is based on `outer`.

Examples

```
x <- 1:9; names(x) <- x
# Multiplication & Power Tables
x %o% x
y <- 2:8; names(y) <- paste(y,":",sep="")
outer(y, x, "^")

outer(month.abb, 1999:2003, FUN = "paste")

## three way multiplication table:
x %o% x %o% y[1:3]
```

`p.adjust` *Adjust p-values for multiple comparisons*

Description

Given a set of p-values, returns p-values adjusted using one of several methods.

Usage

```
p.adjust(p, method=p.adjust.methods, n=length(p))

p.adjust.methods
```

Arguments

`p`	vector of p-values
`method`	correction method
`n`	number of comparisons

Details

The adjustment methods include the Bonferroni correction (`"bonferroni"`) in which the p-values are multiplied by the number of comparisons. Four less conservative corrections are also included by Holm (1979) (`"holm"`), Hochberg (1988) (`"hochberg"`), Hommel (1988) (`"hommel"`) and Benjamini & Hochberg (1995) (`"fdr"`), respectively. A pass-through option (`"none"`) is also included. The set of methods are contained in the `p.adjust.methods` vector for the benefit of methods that need to have the method as an option and pass it on to `p.adjust`.

The first four methods are designed to give strong control of the family wise error rate. There seems no reason to use the unmodified Bonferroni correction because it is dominated by Holm's method, which is also valid under arbitrary assumptions.

Hochberg's and Hommel's methods are valid when the hypothesis tests are independent or when they are non-negatively associated (Sarkar, 1998; Sarkar and Chang, 1997). Hommel's method is more powerful than Hochberg's, but the difference is usually small and the Hochberg p-values are faster to compute.

The `"fdr"` method of Benjamini and Hochberg (1995) controls the false discovery rate, the expected proportion of false discoveries amongst the rejected hypotheses. The false discovery rate is a less stringent condition than the family wise error rate, so Benjamini and Hochberg's method is more powerful than the other methods.

Value

A vector of corrected p-values.

References

Benjamini, Y., and Hochberg, Y. (1995). Controlling the false discovery rate: a practical and powerful approach to multiple testing. *Journal of the Royal Statistical Society Series* B, **57**, 289–300.

Holm, S. (1979). A simple sequentially rejective multiple test procedure. *Scandinavian Journal of Statistics*, **6**, 65–70.

Hommel, G. (1988). A stagewise rejective multiple test procedure based on a modified Bonferroni test. *Biometrika*, **75**, 383–386.

Hochberg, Y. (1988). A sharper Bonferroni procedure for multiple tests of significance. *Biometrika*, **75**, 800–803.

Shaffer, J. P. (1995). Multiple hypothesis testing. *Annual Review of Psychology*, **46**, 561–576. (An excellent review of the area.)

Sarkar, S. (1998). Some probability inequalities for ordered MTP2 random variables: a proof of Simes conjecture. *Annals of Statistics*, **26**, 494–504.

Sarkar, S., and Chang, C. K. (1997). Simes' method for multiple hypothesis testing with positively dependent test statistics. *Journal of the American Statistical Association*, **92**, 1601–1608.

Wright, S. P. (1992). Adjusted P-values for simultaneous inference. *Biometrics*, **48**, 1005–1013. (Explains the adjusted P-value approach.)

See Also

`pairwise.*` functions in the **ctest** package, such as `pairwise.t.test`.

Examples

```
x <- rnorm(50, m=c(rep(0,25),rep(3,25)))
p <- 2*pnorm( -abs(x))
round(p, 3)
round(p.adjust(p), 3)
round(p.adjust(p,"bonferroni"), 3)
round(p.adjust(p,"fdr"), 3)
```

package.contents *Package Contents and Description*

Description

Parses and returns the 'CONTENTS' and 'DESCRIPTION' file of a package.

Usage

```
package.contents(pkg, lib.loc = NULL)
package.description(pkg, lib.loc = NULL, fields = NULL)
```

Arguments

`pkg`	a character string with the package name.
`lib.loc`	a character vector describing the location of R library trees to search through, or `NULL`. The default value of `NULL` corresponds to all libraries currently known.
`fields`	a character vector giving the tags of fields to return (if other fields occur in the file they are ignored).

Value

`package.contents` returns `NA` if there is no 'CONTENTS' file for the given package; otherwise, a character matrix with column names `c("Entry", "Keywords", "Description")` and rows giving the corresponding entries in the CONTENTS data base for each Rd file in the package.

If a 'DESCRIPTION' for the given package is found and can successfully be read, `package.description` returns a named character vector with the values of the (given) fields as elements and the tags as names. If not, it returns a named vector of `NA`s with the field tags as names if `fields` is not null, and `NA` otherwise.

See Also

`read.dcf`

Examples

```
package.contents("mva")
package.contents("mva")[, c("Entry", "Description")]
```

```
package.description("ts")
package.description("ts")[c("Package", "Version")]
## NOTE: No subscripting using '$' or abbreviated field
## tags!
```

`package.dependencies` *Check Package Dependencies*

Description

Parses and checks the dependencies of a package against the currently installed version of R [and other packages].

Usage

```
package.dependencies(x, check=FALSE)
```

Arguments

`x`	A matrix of package descriptions as returned by `CRAN.packages`.
`check`	If `TRUE`, return logical vector of check results. If `FALSE`, return parsed list of dependencies.

Details

Currently we only check if the package conforms with the currently running version of R. In the future we might add checks for inter-package dependencies.

See Also

`update.packages`

package.skeleton *Create a skeleton for a new package*

Description

package.skeleton automates some of the setup for a new package. It creates directories, saves functions and data to appropriate places, and creates skeleton help files and 'README' files describing further steps in packaging.

Usage

```
package.skeleton(name="anRpackage", list,
                 environment=.GlobalEnv,
                 path=".", force=FALSE)
```

Arguments

name	directory name for your package
list	vector of names of R objects to put in the package
environment	if list is omitted, the contents of this environment are packaged
path	path to put the package directories in
force	If FALSE will not overwrite an existing directory

Value

used for its side-effects.

References

Read the *Writing R Extensions* manual for more details

See Also

install.packages

Examples

```
f<-function(x,y) x+y
g<-function(x,y) x-y
d<-data.frame(a=1,b=2)
e<-rnorm(1000)
package.skeleton(list=c("f","g","d","e"),
                 name="AnExample")
```

packageStatus *Package Management Tools*

Description

Summarize information about installed packages and packages available at various repositories, and automatically upgrade outdated packages. These tools will replace `update.packages` and friends in the future and are currently work in progress.

Usage

```
packageStatus(lib.loc = NULL,
              repositories = getOption("repositories"))

## S3 method for class 'packageStatus':
summary(object, ...)

## S3 method for class 'packageStatus':
update(object, lib.loc = levels(object$inst$LibPath),
       repositories = levels(object$avail$Repository), ...)

## S3 method for class 'packageStatus':
upgrade(object, ask = TRUE, ...)
```

Arguments

`lib.loc`	a character vector describing the location of R library trees to search through, or `NULL`. The default value of `NULL` corresponds to all libraries currently known.
`repositories`	a character vector of URLs describing the location of R package repositories on the Internet or on the local machine.
`object`	return value of `packageStatus`.
`ask`	if `TRUE`, the user is prompted which packages should be upgraded and which not.
`...`	currently not used.

Examples

```
x <- packageStatus()
print(x)
```

```
summary(x)
upgrade(x)
x <- update(x)
print(x)
```

page *Invoke a Pager on an R Object*

Description

Displays a representation of the object named by `x` in a pager.

Usage

```
page(x, method = c("dput", "print"), ...)
```

Arguments

`x`	the name of an R object.
`method`	The default method is to dump the object *via* `dput`. An alternative is to print to a file.
`...`	additional arguments for `file.show`. Intended for setting `pager` as `title` and `delete.file` are already used.

See Also

`file.show`, `edit`, `fix`.

To go to a new page when graphing, see `frame`.

Paren *Parentheses and Braces*

Description

Open parenthesis, (, and open brace, {, are `.Primitive` functions in R. Effectively, (is semantically equivalent to the identity `function(x) x`, whereas { is slightly more interesting, see examples.

Usage

```
( ... )

{ ... }
```

References

Becker, R. A., Chambers, J. M. and Wilks, A. R. (1988) *The New S Language*. Wadsworth & Brooks/Cole.

See Also

`if`, `return`, etc for other objects used in the R language itself.

`Syntax` for operator precedence.

Examples

```
f <- get("(")
e <- expression(3 + 2 * 4)
f(e) == e              # TRUE

do <- get("{")
do(x <- 3, y <- 2*x-3, 6-x-y); x; y
```

parse *Parse Expressions*

Description

`parse` returns the parsed but unevaluated expressions in a list. Each element of the list is of mode `expression`.

Usage

```
parse(file = "", n = NULL, text = NULL, prompt = "?")
```

Arguments

`file`	a connection, or a character string giving the name of a file or a URL to read the expressions from. If `file` is `""` and `text` is missing or `NULL` then input is taken from the console.
`n`	the number of statements to parse. If `n` is negative the file is parsed in its entirety.
`text`	character vector. The text to parse. Elements are treated as if they were lines of a file.
`prompt`	the prompt to print when parsing from the keyboard. `NULL` means to use R's prompt, `getOption("prompt")`.

Details

All versions of R accept input from a connection with end of line marked by LF (as used on Unix), CRLF (as used on DOS/Windows) or CR (as used on classic MacOS). The final line can be incomplete, that is missing the final EOL marker.

See `source` for the limits on the size of functions that can be parsed (by default).

References

Becker, R. A., Chambers, J. M. and Wilks, A. R. (1988) *The New S Language.* Wadsworth & Brooks/Cole.

See Also

`scan`, `source`, `eval`, `deparse`.

Examples

```
cat("x <- c(1,4)\n  x ^ 3 -10 ; outer(1:7,5:9)\n",
    file="xyz.Rdmped")
# parse 3 statements from the file "xyz.Rdmped"
parse(file = "xyz.Rdmped", n = 3)
unlink("xyz.Rdmped")
```

paste *Concatenate Strings*

Description

Concatenate vectors after converting to character.

Usage

```
paste(..., sep = " ", collapse = NULL)
```

Arguments

`...`	one or more R objects, to be coerced to character vectors.
`sep`	a character string to separate the terms.
`collapse`	an optional character string to separate the results.

Details

`paste` converts its arguments to character strings, and concatenates them (separating them by the string given by `sep`). If the arguments are vectors, they are concatenated term-by-term to give a character vector result.

If a value is specified for `collapse`, the values in the result are then concatenated into a single string, with the elements being separated by the value of `collapse`.

Value

A character vector of the concatenated values. This will be of length zero if all the objects are of length zero, unless `collapse` is non-NULL, in which case it is a single empty string.

References

Becker, R. A., Chambers, J. M. and Wilks, A. R. (1988) *The New S Language*. Wadsworth & Brooks/Cole.

See Also

String manipulation with `as.character`, `substr`, `nchar`, `strsplit`; further, `cat` which concatenates and writes to a file, and `sprintf` for C like string construction.

Examples

```
paste(1:12) # same as as.character(1:12)
paste("A", 1:6, sep = "")
paste("Today is", date())
```

path.expand *Expand File Paths*

Description

Expand a path name, for example by replacing a leading tilde by the user's home directory (if defined on that platform).

Usage

```
path.expand(path)
```

Arguments

path	character vector containing one or more path names.

Details

On some Unix versions, a leading `~user` will expand to the home directory of `user`, but not on Unix versions without `readline` installed.

See Also

`basename`

Examples

```
path.expand("~/foo")
```

PkgUtils *Utilities for Building and Checking Add-on Packages*

Description

Utilities for checking whether the sources of an R add-on package work correctly, and for building a source or binary package from them.

Usage

```
R CMD build [options] pkgdirs
R CMD check [options] pkgdirs
```

Arguments

`pkgdirs`	a list of names of directories with sources of R add-on packages.
`options`	further options to control the processing, or for obtaining information about usage and version of the utility.

Details

`R CMD check` checks R add-on packages from their sources, performing a wide variety of diagnostic checks.

`R CMD build` builds R source or binary packages from their sources. It will create index files in the sources if necessary, so it is often helpful to run `build` before `check`.

Use `R CMD foo --help` to obtain usage information on utility `foo`.

Several of the options to `build --binary` are passed to `INSTALL` so consult its help for the details.

See Also

The chapter "Processing Rd format" in "Writing R Extensions" (see the '`doc/manual`' subdirectory of the R source tree).

`INSTALL` is called by `build --binary`.

pmatch *Partial String Matching*

Description

pmatch seeks matches for the elements of its first argument among those of its second.

Usage

```
pmatch(x, table, nomatch = NA, duplicates.ok = FALSE)
```

Arguments

`x`	the values to be matched.
`table`	the values to be matched against.
`nomatch`	the value returned at non-matching or multiply partially matching positions.
`duplicates.ok`	should elements in `table` be used more than once?

Details

The behaviour differs by the value of `duplicates.ok`. Consider first the case if this is true. First exact matches are considered, and the positions of the first exact matches are recorded. Then unique partial matches are considered, and if found recorded. (A partial match occurs if the whole of the element of `x` matches the beginning of the element of `table`.) Finally, all remaining elements of `x` are regarded as unmatched. In addition, an empty string can match nothing, not even an exact match to an empty string. This is the appropriate behaviour for partial matching of character indices, for example.

If `duplicates.ok` is `FALSE`, values of `table` once matched are excluded from the search for subsequent matches. This behaviour is equivalent to the R algorithm for argument matching, except for the consideration of empty strings (which in argument matching are matched after exact and partial matching to any remaining arguments).

`charmatch` is similar to `pmatch` with `duplicates.ok` true, the differences being that it differentiates between no match and an ambiguous partial match, it does match empty strings, and it does not allow multiple exact matches.

Value

A numeric vector of integers (including `NA` if `nomatch = NA`) of the same length as `x`, giving the indices of the elements in `table` which matched, or `nomatch`.

References

Becker, R. A., Chambers, J. M. and Wilks, A. R. (1988) *The New S Language.* Wadsworth & Brooks/Cole.

Chambers, J. M. (1998) *Programming with Data. A Guide to the S Language.* Springer.

See Also

`match`, `charmatch` and `match.arg`, `match.fun`, `match.call`, for function argument matching etc., `grep` etc for more general (regexp) matching of strings.

Examples

```
pmatch("", "")                              # returns NA
pmatch("m",   c("mean", "median", "mode")) # returns NA
pmatch("med", c("mean", "median", "mode")) # returns 2

pmatch(c("", "ab", "ab"), c("abc", "ab"), dup=FALSE)
pmatch(c("", "ab", "ab"), c("abc", "ab"), dup=TRUE)
## compare
charmatch(c("", "ab", "ab"), c("abc", "ab"))
```

`pos.to.env` *Convert Positions in the Search Path to Environments*

Description

Returns the environment at a specified position in the search path.

Usage

```
pos.to.env(x)
```

Arguments

`x`	an integer between 1 and `length(search())`, the length of the search path.

Details

Several R functions for manipulating objects in environments (such as `get` and `ls`) allow specifying environments via corresponding positions in the search path. `pos.to.env` is a convenience function for programmers which converts these positions to corresponding environments; users will typically have no need for it.

Examples

```
pos.to.env(1) # R_GlobalEnv
# the next returns NULL, which is how package:base is
# represented.
pos.to.env(length(search()))
```

predict *Model Predictions*

Description

`predict` is a generic function for predictions from the results of various model fitting functions. The function invokes particular *methods* which depend on the `class` of the first argument.

The function `predict.lm` makes predictions based on the results produced by `lm`.

Usage

```
predict (object, ...)
```

Arguments

`object`	a model object for which prediction is desired.
`...`	additional arguments affecting the predictions produced.

Value

The form of the value returned by `predict` depends on the class of its argument. See the documentation of the particular methods for details of what is produced by that method.

References

Chambers, J. M. and Hastie, T. J. (1992) *Statistical Models in S.* Wadsworth & Brooks/Cole.

See Also

`predict.lm`.

Examples

```
## All the "predict" methods visible in your current
## search() path. NB most of the methods in the base
## packages are hidden.
for(fn in methods("predict"))
   try(cat(fn,":\n\t",deparse(args(get(fn))),"\n"),
       silent = TRUE)
```

`Primitive` *Call a "Primitive" Internal Function*

Description

`.Primitive` returns an entry point to a "primitive" (internally implemented) function.

The advantage of `.Primitive` over `.Internal` functions is the potential efficiency of argument passing.

Usage

```
.Primitive(name)
```

Arguments

`name`	name of the R function.

See Also

`.Internal`.

Examples

```
mysqrt <- .Primitive("sqrt")
c
.Internal # this one *must* be primitive!
get("if") # just 'if' or 'print(if)' are not valid syntax.
```

print *Print Values*

Description

print prints its argument and returns it *invisibly* (via invisible(x)). It is a generic function which means that new printing methods can be easily added for new classes.

Usage

```
print(x, ...)

## S3 method for class 'factor':
print(x, quote = FALSE, max.levels = NULL,
      width = getOption("width"), ...)

## S3 method for class 'table':
print(x, digits = getOption("digits"), quote = FALSE,
       na.print = "", zero.print = "0", justify = "none",
       ...)
```

Arguments

x	an object used to select a method.
...	further arguments passed to or from other methods.
quote	logical, indicating whether or not strings should be printed with surrounding quotes.
max.levels	integer, indicating how many levels should be printed for a factor; if 0, no extra "Levels" line will be printed. The default, NULL, entails choosing max.levels such that the levels print on one line of width width.
width	only used when max.levels is NULL, see above.
digits	minimal number of *significant* digits, see print.default.
na.print	character string (or NULL) indicating NA values in printed output, see print.default.
zero.print	character specifying how zeros (0) should be printed; for sparse tables, using "." can produce stronger results.

`justify` character indicating if strings should left- or right-justified or left alone, passed to `format`.

Details

The default method, `print.default` has its own help page. Use `methods("print")` to get all the methods for the `print` generic.

`print.factor` allows some customization and is used for printing `ordered` factors as well.

`print.table` for printing `table`s allows other customization.

See `noquote` as an example of a class whose main purpose is a specific `print` method.

References

Chambers, J. M. and Hastie, T. J. (1992) *Statistical Models in S.* Wadsworth & Brooks/Cole.

See Also

The default method `print.default`, and help for the methods above; further `options`, `noquote`.

For more customizable (but cumbersome) printing, see `cat`, `format` or also `write`.

Examples

```
ts(1:20) # print is the "Default function"
         # therefore print.ts(.) is called
rr <- for(i in 1:3) print(1:i)
rr

## Printing of factors
data(attenu)
attenu$station ## 117 levels -> `max.levels'
               ## depending on width

data(esoph) ## ordered : levels  "l1 < l2 < .."
esoph$agegp[1:12]
esoph$alcgp[1:12]

## Printing of sparse (contingency) tables
set.seed(521)
```

```
t1 <- round(abs(rt(200, df=1.8)))
t2 <- round(abs(rt(200, df=1.4)))
table(t1,t2) # simple
print(table(t1,t2), zero.print = ".") # nicer to read
```

print.data.frame *Printing Data Frames*

Description

Print a data frame.

Usage

```
## S3 method for class 'data.frame':
print(x, ..., digits = NULL, quote = FALSE, right = TRUE)
```

Arguments

`x`	object of class `data.frame`.
`...`	optional arguments to `print` or `plot` methods.
`digits`	the minimum number of significant digits to be used.
`quote`	logical, indicating whether or not entries should be printed with surrounding quotes.
`right`	logical, indicating whether or not strings should be right-aligned. The default is left-alignment.

Details

This calls `format` which formats the data frame column-by-column, then converts to a character matrix and dispatches to the `print` method for matrices.

When `quote = TRUE` only the entries are quoted not the row names nor the column names.

See Also

`data.frame`.

`print.default` *Default Printing*

Description

`print.default` is the *default* method of the generic `print` function which prints its argument. `print.matrix` is currently identical, but was not prior to `1.7.0`.

Usage

```
## Default S3 method:
print(x, digits = NULL, quote = TRUE, na.print = NULL,
      print.gap = NULL, right = FALSE, ...)
```

Arguments

`x`	the object to be printed.
`digits`	a non-null value for `digits` specifies the minimum number of significant digits to be printed in values. If `digits` is `NULL`, the value of `digits` set by `options` is used.
`quote`	logical, indicating whether or not strings (`characters`) should be printed with surrounding quotes.
`na.print`	a character string which is used to indicate `NA` values in printed output, or `NULL` (see Details)
`print.gap`	an integer, giving the spacing between adjacent columns in printed matrices and arrays, or `NULL` meaning 1.
`right`	logical, indicating whether or not strings should be right-aligned. The default is left-alignment.
`...`	further arguments to be passed to or from other methods. They are ignored in these functions.

Details

Prior to R 1.7.0, `print.matrix` did not print attributes and did not have a `digits` argument.

The default for printing `NA`s is to print `NA` (without quotes) unless this is a character `NA` *and* `quote = FALSE`, when `<NA>` is printed.

The same number of decimal places is used throughout a vector, This means that `digits` specifies the minimum number of significant digits to be used, and that at least one entry will be printed with that minimum number.

As from R 1.7.0 attributes are printed respecting their class(es), using the values of `digits` to `print.default`, but using the default values (for the methods called) of the other arguments.

When the **methods** package is attached, `print` will call `show` for methods with formal classes if called with no optional arguments.

See Also

The generic `print`, `options`. The `"noquote"` class and print method.

Examples

```
pi
print(pi, digits = 16)
LETTERS[1:16]
print(LETTERS, quote = FALSE)
```

printCoefmat *Print Coefficient Matrices*

Description

Utility function to be used in "higher level" `print` methods, such as `print.summary.lm`, `print.summary.glm` and `print.anova`. The goal is to provide a flexible interface with smart defaults such that often, only `x` needs to be specified.

Usage

```
printCoefmat(x, digits=max(3, getOption("digits") - 2),
  signif.stars = getOption("show.signif.stars"),
  dig.tst = max(1, min(5, digits - 1)),
  cs.ind = 1:k, tst.ind = k + 1, zap.ind = integer(0),
  P.values = NULL,
  has.Pvalue = nc >= 4 &&
    substr(colnames(x)[nc],1,3) == "Pr(",
  eps.Pvalue = .Machine$double.eps,
  na.print = "NA", ...)
```

Arguments

`x`	a numeric matrix like object, to be printed.
`digits`	minimum number of significant digits to be used for most numbers.
`signif.stars`	logical; if `TRUE`, P-values are additionally encoded visually as "significance stars" in order to help scanning of long coefficient tables. It defaults to the `show.signif.stars` slot of `options`.
`dig.tst`	minimum number of significant digits for the test statistics, see `tst.ind`.
`cs.ind`	indices (integer) of column numbers which are (like) coefficients and standard errors to be formatted together.
`tst.ind`	indices (integer) of column numbers for test statistics.
`zap.ind`	indices (integer) of column numbers which should be formatted by `zapsmall`, i.e., by "zapping" values close to 0.

`P.values` logical or `NULL`; if `TRUE`, the last column of `x` is formatted by `format.pval` as P values. If `P.values = NULL`, the default, it is set to `TRUE` only if `link{options}("show.coef.Pvalue")` is `TRUE` *and* `x` has at least 4 columns *and* the last column name of `x` starts with `"Pr("`.

`has.Pvalue` logical; if `TRUE`, the last column of `x` contains P values; in that case, it is printed *if and only if* `P.values` (above) is true.

`eps.Pvalue` number,..

`na.print` a character string to code `NA` values in printed output.

`...` further arguments for `print`.

Value

Invisibly returns its argument, `x`.

Author(s)

Martin Maechler

See Also

`print.summary.lm`, `format.pval`, `format`.

Examples

```
cmat <- cbind(rnorm(3, 10), sqrt(rchisq(3, 12)))
cmat <- cbind(cmat, cmat[,1]/cmat[,2])
cmat <- cbind(cmat, 2*pnorm(-cmat[,3]))
colnames(cmat) <- c("Estimate", "Std.Err",
                    "Z value", "Pr(>z)")
printCoefmat(cmat[,1:3])
printCoefmat(cmat)
options(show.coef.Pvalues = FALSE)
printCoefmat(cmat, digits=2)
printCoefmat(cmat, digits=2, P.values = TRUE)
options(show.coef.Pvalues = TRUE) # revert
```

`prmatrix` *Print Matrices, Old-style*

Description

An earlier method for printing matrices, provided for S compatibility.

Usage

```
prmatrix(x, rowlab =, collab =,
         quote = TRUE, right = FALSE, na.print = NULL, ...)
```

Arguments

`x`	numeric or character matrix.
`rowlab,collab`	(optional) character vectors giving row or column names respectively. By default, these are taken from `dimnames(x)`.
`quote`	logical; if `TRUE` and `x` is of mode `"character"`, *quotes* (`"`) are used.
`right`	if `TRUE` and `x` is of mode `"character"`, the output columns are *right*-justified.
`na.print`	how `NA`s are printed. If this is non-null, its value is used to represent `NA`.
`...`	arguments for `print` methods.

Details

`prmatrix` is an earlier form of `print.matrix`, and is very similar to the S function of the same name.

Value

Invisibly returns its argument, `x`.

References

Becker, R. A., Chambers, J. M. and Wilks, A. R. (1988) *The New S Language.* Wadsworth & Brooks/Cole.

See Also

`print.default`, and other `print` methods.

Examples

```
prmatrix(m6 <- diag(6), row = rep("",6), coll=rep("",6))

chm <- matrix(scan(system.file("help", "AnIndex",
                               package = "eda"),
                   what = ""), , 2, byrow = TRUE)
chm  # uses print.matrix()
prmatrix(chm, collab = paste("Column",1:3), right=TRUE,
         quote=FALSE)
```

`proc.time` *Running Time of R*

Description

`proc.time` determines how much time (in seconds) the currently running R process already consumed.

Usage

```
proc.time()
```

Value

A numeric vector of length 5, containing the user, system, and total elapsed times for the currently running R process, and the cumulative sum of user and system times of any child processes spawned by it.

The resolution of the times will be system-specific; it is common for them to be recorded to of the order of 1/100 second, and elapsed time is rounded to the nearest 1/100.

It is most useful for "timing" the evaluation of R expressions, which can be done conveniently with `system.time`.

Note

It is possible to compile R without support for `proc.time`, when the function will not exist.

References

Becker, R. A., Chambers, J. M. and Wilks, A. R. (1988) *The New S Language*. Wadsworth & Brooks/Cole.

See Also

`system.time` for timing a valid R expression, `gc.time` for how much of the time was spent in garbage collection.

Examples

```
## a way to time an R expression: system.time is preferred
ptm <- proc.time()
for (i in 1:50) mad(runif(500))
proc.time() - ptm
```

prompt *Produce Prototype of an R Documentation File*

Description

Facilitate the constructing of files documenting R objects.

Usage

```
prompt(object, filename = NULL, name = NULL, ...)
## Default S3 method:
prompt(object, filename = NULL, name = NULL,
       force.function = FALSE, ...)
## S3 method for class 'data.frame':
prompt(object, filename = NULL, name = NULL, ...)
```

Arguments

`object` an R object, typically a function for the default method.

`filename` usually, a connection or a character string giving the name of the file to which the documentation shell should be written. The default corresponds to a file whose name is `name` followed by `".Rd"`. Can also be `NA` (see below).

`name` a character string specifying the name of the object.

`force.function` a logical. If `TRUE`, treat `object` as function in any case.

`...` further arguments passed to or from other methods.

Details

Unless `filename` is `NA`, a documentation shell for `object` is written to the file specified by `filename`, and a message about this is given. For function objects, this shell contains the proper function and argument names. R documentation files thus created still need to be edited and moved into the 'man' subdirectory of the package containing the object to be documented.

If `filename` is `NA`, a list-style representation of the documentation shell is created and returned. Writing the shell to a file amounts to `cat(unlist(x), file = filename, sep = "\n")`, where `x` is the list-style representation.

When `prompt` is used in `for` loops or scripts, the explicit `name` specification will be useful.

Value

If `filename` is `NA`, a list-style representation of the documentation shell. Otherwise, the name of the file written to is returned invisibly.

Warning

Currently, calling `prompt` on a non-function object assumes that the object is in fact a data set and hence documents it as such. This may change in future versions of R. Use `promptData` to create documentation skeletons for data sets.

Note

The documentation file produced by `prompt.data.frame` does not have the same format as many of the data frame documentation files in the **base** package. We are trying to settle on a preferred format for the documentation.

Author(s)

Douglas Bates for `prompt.data.frame`

References

Becker, R. A., Chambers, J. M. and Wilks, A. R. (1988) *The New S Language*. Wadsworth & Brooks/Cole.

See Also

`promptData`, `help` and the chapter on "Writing R documentation" in "Writing R Extensions" (see the '`doc/manual`' subdirectory of the R source tree).

To prompt the user for input, see `readline`.

Examples

```
prompt(plot.default)
prompt(interactive, force.function = TRUE)
unlink("plot.default.Rd")
unlink("interactive.Rd")

data(women) # data.frame
prompt(women)
unlink("women.Rd")

data(sunspots) # non-data.frame data
prompt(sunspots)
unlink("sunspots.Rd")
```

promptData *Generate a Shell for Documentation of Data Sets*

Description

Generates a shell of documentation for a data set.

Usage

```
promptData(object, filename = NULL, name = NULL)
```

Arguments

`object`	an R object to be documented as a data set.
`filename`	usually, a connection or a character string giving the name of the file to which the documentation shell should be written. The default corresponds to a file whose name is `name` followed by `".Rd"`. Can also be `NA` (see below).
`name`	a character string specifying the name of the object.

Details

Unless `filename` is `NA`, a documentation shell for `object` is written to the file specified by `filename`, and a message about this is given.

If `filename` is `NA`, a list-style representation of the documentation shell is created and returned. Writing the shell to a file amounts to `cat(unlist(x), file = filename, sep = "\n")`, where `x` is the list-style representation.

Currently, only data frames are handled explicitly by the code.

Value

If `filename` is `NA`, a list-style representation of the documentation shell. Otherwise, the name of the file written to is returned invisibly.

Warning

This function is still experimental. Both interface and value might change in future versions. In particular, it may be preferable to use a character string naming the data set and optionally a specification of where to look for it instead of using `object`/`name` as we currently do.

This would be different from `prompt`, but consistent with other prompt-style functions in package **methods**, and also allow prompting for data set documentation without explicitly having to load the data set.

See Also

`prompt`

Examples

```
data(sunspots)
promptData(sunspots)
unlink("sunspots.Rd")
```

prop.table *Express table entries as fraction of marginal table*

Description

This is really `sweep(x, margin, margin.table(x, margin), "/")` for newbies, except that if `margin` has length zero, then one gets `x/sum(x)`.

Usage

```
prop.table(x, margin=NULL)
```

Arguments

`x`	table
`margin`	index, or vector of indices to generate margin for

Value

Table like `x` expressed relative to `margin`

Author(s)

Peter Dalgaard

See Also

`margin.table`

Examples

```
m<-matrix(1:4,2)
m
prop.table(m,1)
```

pushBack *Push Text Back on to a Connection*

Description

Functions to push back text lines onto a connection, and to enquire how many lines are currently pushed back.

Usage

```
pushBack(data, connection, newLine = TRUE)
pushBackLength(connection)
```

Arguments

`data`	a character vector.
`connection`	A connection.
`newLine`	logical. If true, a newline is appended to each string pushed back.

Details

Several character strings can be pushed back on one or more occasions. The occasions form a stack, so the first line to be retrieved will be the first string from the last call to `pushBack`. Lines which are pushed back are read prior to the normal input from the connection, by the normal text-reading functions such as `readLines` and `scan`.

Pushback is only allowed for readable connections.

Not all uses of connections respect pushbacks, in particular the input connection is still wired directly, so for example parsing commands from the console and `scan("")` ignore pushbacks on `stdin`.

Value

`pushBack` returns nothing.

`pushBackLength` returns number of lines currently pushed back.

See Also

`connections`, `readLines`.

Examples

```
zz <- textConnection(LETTERS)
readLines(zz, 2)
pushBack(c("aa", "bb"), zz)
pushBackLength(zz)
readLines(zz, 1)
pushBackLength(zz)
readLines(zz, 1)
readLines(zz, 1)
close(zz)
```

quantile *Sample Quantiles*

Description

The generic function `quantile` produces sample quantiles corresponding to the given probabilities. The smallest observation corresponds to a probability of 0 and the largest to a probability of 1.

Usage

```
quantile(x, ...)

## Default S3 method:
quantile(x, probs = seq(0, 1, 0.25), na.rm = FALSE,
         names = TRUE, ...)
```

Arguments

`x`	numeric vectors whose sample quantiles are wanted.
`probs`	numeric vector with values in $[0, 1]$.
`na.rm`	logical; if true, any `NA` and `NaN`'s are removed from `x` before the quantiles are computed.
`names`	logical; if true, the result has a `names` attribute. Set to `FALSE` for speedup with many `probs`.
`...`	further arguments passed to or from other methods.

Details

A vector of length `length(probs)` is returned; if `names = TRUE`, it has a `names` attribute.

`quantile(x,p)` as a function of `p` linearly interpolates the points $((i-1)/(n-1), ox[i])$, where `ox <- sort(x)` and `n <- length(x)`.

This gives `quantile(x, p) == (1-f)*ox[i] + f*ox[i+1]`, where `r <- 1 + (n-1)*p`, `i <- floor(r)`, `f <- r - i` *and* `ox[n+1] := ox[n]`.

`NA` and `NaN` values in `probs` are propagated to the result.

References

Becker, R. A., Chambers, J. M. and Wilks, A. R. (1988) *The New S Language*. Wadsworth & Brooks/Cole.

See Also

`ecdf` (in the **stepfun** package) for empirical distributions of which `quantile` is the "inverse"; `boxplot.stats` and `fivenum` for computing "versions" of quartiles, etc.

Examples

```
quantile(x <- rnorm(1001)) # Extremes & Quartiles by default
quantile(x,  probs=c(.1,.5,1,2,5,10,50, NA)/100)
```

quit *Terminate an R Session*

Description

The function `quit` or its alias `q` terminate the current R session.

Usage

```
quit(save = "default", status = 0, runLast = TRUE)
   q(save = "default", status = 0, runLast = TRUE)
.Last <- function(x) { ...... }
```

Arguments

`save`	a character string indicating whether the environment (workspace) should be saved, one of `"no"`, `"yes"`, `"ask"` or `"default"`.
`status`	the (numerical) error status to be returned to the operating system, where relevant. Conventionally `0` indicates successful completion.
`runLast`	should `.Last()` be executed?

Details

`save` must be one of `"no"`, `"yes"`, `"ask"` or `"default"`. In the first case the workspace is not saved, in the second it is saved and in the third the user is prompted and can also decide *not* to quit. The default is to ask in interactive use but may be overridden by command-line arguments (which must be supplied in non-interactive use).

Immediately *before* terminating, the function `.Last()` is executed if it exists and `runLast` is true. If in interactive use there are errors in the `.Last` function, control will be returned to the command prompt, so do test the function thoroughly.

Some error statuses are used by R itself. The default error handler for non-interactive effectively calls `q("no", 1, FALSE)` and returns error code 1. Error status 2 is used for catastrophic failure, and other small numbers are used by specific ports for initialization failures. It is recommended that users choose statuses of 10 or more.

Valid values of `status` are system-dependent, but `0:255` are normally valid.

References

Becker, R. A., Chambers, J. M. and Wilks, A. R. (1988) *The New S Language.* Wadsworth & Brooks/Cole.

See Also

`.First` for setting things on startup.

Examples

```
## Unix-flavour example
.Last <- function() {
  cat("Now sending PostScript graphics to the printer:\n")
  system("lpr Rplots.ps")
  cat("bye bye...\n")
}
quit("yes")
```

`R.home` *Return the R Home Directory*

Description

Return the R home directory.

Usage

```
R.home()
```

Value

A character string giving the current home directory.

R.Version *Version Information*

Description

`R.Version()` provides detailed information about the version of R running.
`R.version` is a variable (a `list`) holding this information (and `version` is a copy of it for S compatibility), whereas `R.version.string` is a simple `character` string, useful for plotting, etc.

Usage

```
R.Version()
R.version
R.version.string
```

Value

`R.Version` returns a list with components

`platform`	the platform for which R was built. Under Unix, a string of the form CPU-VENDOR-OS, as determined by the configure script. E.g, `"i386-pc-gnu"`.
`arch`	the architecture (CPU) R was built on/for.
`os`	the underlying operating system
`system`	CPU and OS.
`status`	the status of the version (e.g., `"Alpha"`)
`status.rev`	the status revision level
`major`	the major version number
`minor`	the minor version number
`year`	the year the version was released
`month`	the month the version was released
`day`	the day the version was released
`language`	always `"R"`.

Note

Do *not* use `R.version$os` to test the platform the code is running on: use `.Platform$OS.type` instead. Slightly different versions of the OS may report different values of `R.version$os`, as may different versions of R.

See Also

`.Platform`.

Examples

```
R.version$os # to check how lucky you are ...
plot(0) # any plot
# a useful bottom-right note
mtext(R.version.string, side=1,line=4,adj=1)
```

`rank` *Sample Ranks*

Description

Returns the sample ranks of the values in a numeric vector. Ties, i.e., equal values, result in ranks being averaged, by default.

Usage

```
rank(x, na.last = TRUE,
     ties.method = c("average", "first", "random"))
```

Arguments

`x`	a numeric vector.
`na.last`	for controlling the treatment of `NA`s. If `TRUE`, missing values in the data are put last; if `FALSE`, they are put first; if `NA`, they are removed; if `"keep"` they are kept.
`ties.method`	a character string specifying how ties are treated, see below; can be abbreviated.

Details

If all components are different, the ranks are well defined, with values in `1:n` where `n <- length(x)` and we assume no `NA`s for the moment. Otherwise, with some values equal, called 'ties', the argument `ties.method` determines the result at the corresponding indices. The `"first"` method results in a permutation with increasing values at each index set of ties. The `"random"` method puts these in random order whereas the default, `"average"`, replaces them by their mean.

References

Becker, R. A., Chambers, J. M. and Wilks, A. R. (1988) *The New S Language.* Wadsworth & Brooks/Cole.

See Also

`order` and `sort`.

Examples

```
(r1 <- rank(x1 <- c(3, 1, 4, 15, 92)))
x2 <- c(3, 1, 4, 1, 5, 9, 2, 6, 5, 3, 5)
names(x2) <- letters[1:11]
(r2 <- rank(x2)) # ties are averaged

## rank() is "idempotent": rank(rank(x)) == rank(x) :
stopifnot(rank(r1) == r1, rank(r2) == r2)

## ranks without averaging
rank(x2, ties.method= "first")  # first occurrence wins
rank(x2, ties.method= "random") # ties broken at random
rank(x2, ties.method= "random") # and again
```

RdUtils *Utilities for Processing Rd Files*

Description

Utilities for converting files in R documentation (Rd) format to other formats or create indices from them, and for converting documentation in other formats to Rd format.

Usage

```
R CMD Rdconv [options] file
R CMD Rd2dvi [options] files
R CMD Rd2txt [options] file
R CMD Sd2Rd [options] file
```

Arguments

`file`	the path to a file to be processed.
`files`	a list of file names specifying the R documentation sources to use, by either giving the paths to the files, or the path to a directory with the sources of a package.
`options`	further options to control the processing, or for obtaining information about usage and version of the utility.

Details

`Rdconv` converts Rd format to other formats. Currently, plain text, HTML, LaTeX, S version 3 (Sd), and S version 4 (.sgml) formats are supported. It can also extract the examples for run-time testing.

`Rd2dvi` and `Rd2txt` are user-level programs for producing DVI/PDF output or pretty text output from Rd sources.

`Sd2Rd` converts S (version 3 or 4) documentation formats to Rd format.

Use `R CMD foo --help` to obtain usage information on utility `foo`.

Note

Conversion to S version 3/4 formats is rough: there are some `.Rd` constructs for which there is no natural analogue. They are intended as a starting point for hand-tuning.

See Also

The chapter "Processing Rd format" in "Writing R Extensions" (see the 'doc/manual' subdirectory of the R source tree).

`read.00Index` *Read 00Index-style Files*

Description

Read item/description information from 00Index-style files. Such files are description lists rendered in tabular form, and currently used for the object, data and demo indices and 'TITLE' files of add-on packages.

Usage

```
read.00Index(file)
```

Arguments

`file`	the name of a file to read data values from. If the specified file is `""`, then input is taken from the keyboard (in this case input can be terminated by a blank line). Alternatively, `file` can be a `connection`, which will be opened if necessary, and if so closed at the end of the function call.

Value

a character matrix with 2 columns named `"Item"` and `"Description"` which hold the items and descriptions.

See Also

`formatDL` for the inverse operation of creating a 00Index-style file from items and their descriptions.

read.ftable *Manipulate Flat Contingency Tables*

Description

Read, write and coerce "flat" contingency tables.

Usage

```
read.ftable(file, sep = "", quote = "\"",
            row.var.names, col.vars, skip = 0)
write.ftable(x, file = "", quote = TRUE,
             digits = getOption("digits"))

## S3 method for class 'ftable':
as.table(x, ...)
```

Arguments

`file`	either a character string naming a file or a connection which the data are to be read from or written to. `""` indicates input from the console for reading and output to the console for writing.
`sep`	the field separator string. Values on each line of the file are separated by this string.
`quote`	a character string giving the set of quoting characters for `read.ftable`; to disable quoting altogether, use `quote=""`. For `write.table`, a logical indicating whether strings in the data will be surrounded by double quotes.
`row.var.names`	a character vector with the names of the row variables, in case these cannot be determined automatically.
`col.vars`	a list giving the names and levels of the column variables, in case these cannot be determined automatically.
`skip`	the number of lines of the data file to skip before beginning to read data.
`x`	an object of class `"ftable"`.
`digits`	an integer giving the number of significant digits to use for (the cell entries of) `x`.
`...`	further arguments to be passed to or from methods.

Details

`read.ftable` reads in a flat-like contingency table from a file. If the file contains the written representation of a flat table (more precisely, a header with all information on names and levels of column variables, followed by a line with the names of the row variables), no further arguments are needed. Similarly, flat tables with only one column variable the name of which is the only entry in the first line are handled automatically. Other variants can be dealt with by skipping all header information using `skip`, and providing the names of the row variables and the names and levels of the column variable using `row.var.names` and `col.vars`, respectively. See the examples below.

Note that flat tables are characterized by their "ragged" display of row (and maybe also column) labels. If the full grid of levels of the row variables is given, one should instead use `read.table` to read in the data, and create the contingency table from this using `xtabs`.

`write.ftable` writes a flat table to a file, which is useful for generating "pretty" ASCII representations of contingency tables.

`as.table.ftable` converts a contingency table in flat matrix form to one in standard array form. This is a method for the generic function `as.table`.

References

Agresti, A. (1990) *Categorical data analysis.* New York: Wiley.

See Also

`ftable` for more information on flat contingency tables.

Examples

```
## Not in ftable standard format, but o.k.
file <- tempfile()
cat("                Employed\n",
    "State Gender     Yes  No\n",
    "AZ    Male        49  14\n",
    "      Female      58  10\n",
    "NM    Male        52  11\n",
    "      Female      61  13\n",
    file = file)
file.show(file)
ft <- read.ftable(file)
```

```
ft
unlink(file)

## Agresti (1990), page 297, Table 8.16. Almost o.k., but
## misses the name of the row variable.
file <- tempfile()
cat("                              \"Tonsil Size\"\n",
    "            \"Not Enl.\" \"Enl.\" \"Greatly Enl.\"\n",
    "Noncarriers       497     560            269\n",
    "Carriers           19      29             24\n",
    file = file)
file.show(file)
ft <- read.ftable(file, skip = 2,
         row.var.names = "Status",
         col.vars = list("Tonsil Size" =
         c("Not Enl.", "Enl.", "Greatly Enl.")))
ft
unlink(file)
```

read.fwf *Read Fixed Width Format Files*

Description

Read a "table" of fixed width formatted data into a `data.frame`.

Usage

```
read.fwf(file, widths, header = FALSE, sep = "\t",
         as.is = FALSE, skip = 0, row.names, col.names,
         n = -1, ...)
```

Arguments

`file`	the name of the file which the data are to be read from. Alternatively, `file` can be a `connection`, which will be opened if necessary, and if so closed at the end of the function call.
`widths`	integer vector, giving the widths of the fixed-width fields (of one line).
`header`	a logical value indicating whether the file contains the names of the variables as its first line.
`sep`	character; the separator used internally; should be a character that does not occur in the file.
`as.is`	see `read.table`.
`skip`	number of initial lines to skip; see `read.table`.
`row.names`	see `read.table`.
`col.names`	see `read.table`.
`n`	the maximum number of records (lines) to be read, defaulting to no limit.
`...`	further arguments to be passed to `read.table`.

Details

Fields that are of zero-width or are wholly beyond the end of the line in `file` are replaced by `NA`.

Value

A `data.frame` as produced by `read.table` which is called internally.

Author(s)

Brian Ripley for R version: original `Perl` by Kurt Hornik.

See Also

`scan` and `read.table`.

Examples

```
ff <- tempfile()
cat(file=ff, "123456", "987654", sep="\n")
# 1 23 456 \ 9 87 654
read.fwf(ff, width=c(1,2,3))
unlink(ff)
cat(file=ff, "123", "987654", sep="\n")
# 1 NA 23 NA \ 9 NA 87 654
read.fwf(ff, width=c(1,0, 2,3))
unlink(ff)
```

read.socket *Read from or Write to a Socket*

Description

`read.socket` reads a string from the specified socket, `write.socket` writes to the specified socket. There is very little error checking done by either.

Usage

```
read.socket(socket, maxlen=256, loop=FALSE)
write.socket(socket, string)
```

Arguments

`socket`	a socket object
`maxlen`	maximum length of string to read
`loop`	wait for ever if there is nothing to read?
`string`	string to write to socket

Value

`read.socket` returns the string read.

Author(s)

Thomas Lumley

See Also

`close.socket`, `make.socket`

Examples

```
finger <-
  function(user, host="localhost", port=79, print=TRUE)
{
    if (!is.character(user))
        stop("user name must be a string")
    user <- paste(user,"\r\n")
    socket <- make.socket(host, port)
    on.exit(close.socket(socket))
```

```
    write.socket(socket, user)
    output <- character(0)
    repeat{
        ss <- read.socket(socket)
        if (ss == "") break
        output <- paste(output, ss)
    }
    close.socket(socket)
    if (print) cat(output)
    invisible(output)
}
finger("root") ## only works if finger daemon is running
```

read.table *Data Input*

Description

Reads a file in table format and creates a data frame from it, with cases corresponding to lines and variables to fields in the file.

Usage

```
read.table(file, header = FALSE, sep = "", quote = "\"'",
  dec = ".", row.names, col.names, as.is = FALSE,
  na.strings = "NA", colClasses = NA, nrows = -1,
  skip = 0, check.names = TRUE, fill = !blank.lines.skip,
  strip.white = FALSE, blank.lines.skip = TRUE,
  comment.char = "#")

read.csv(file, header = TRUE, sep = ",", quote="\"",
         dec=".", fill = TRUE, ...)

read.csv2(file, header = TRUE, sep = ";", quote="\"",
          dec=",", fill = TRUE, ...)

read.delim(file, header = TRUE, sep = "\t", quote="\"",
           dec=".", fill = TRUE, ...)

read.delim2(file, header = TRUE, sep = "\t", quote="\"",
            dec=",", fill = TRUE, ...)
```

Arguments

`file` the name of the file which the data are to be read from. Each row of the table appears as one line of the file. If it does not contain an *absolute* path, the file name is *relative* to the current working directory, `getwd()`. Tilde-expansion is performed where supported.

Alternatively, `file` can be a `connection`, which will be opened if necessary, and if so closed at the end of the function call. (If `stdin()` is used, the prompts for lines may be somewhat confusing. Terminate input with an EOF signal, `Ctrl-D` on Unix and `Ctrl-Z` on Windows.)

`file` can also be a complete URL.

`header`	a logical value indicating whether the file contains the names of the variables as its first line. If missing, the value is determined from the file format: `header` is set to `TRUE` if and only if the first row contains one fewer field than the number of columns.
`sep`	the field separator character. Values on each line of the file are separated by this character. If `sep = ""` (the default for `read.table`) the separator is "white space", that is one or more spaces, tabs or newlines.
`quote`	the set of quoting characters. To disable quoting altogether, use `quote=""`. See `scan` for the behaviour on quotes embedded in quotes.
`dec`	the character used in the file for decimal points.
`row.names`	a vector of row names. This can be a vector giving the actual row names, or a single number giving the column of the table which contains the row names, or character string giving the name of the table column containing the row names. If there is a header and the first row contains one fewer field than the number of columns, the first column in the input is used for the row names. Otherwise if `row.names` is missing, the rows are numbered. Using `row.names = NULL` forces row numbering.
`col.names`	a vector of optional names for the variables. The default is to use `"V"` followed by the column number.
`as.is`	the default behavior of `read.table` is to convert character variables (which are not converted to logical, numeric or complex) to factors. The variable `as.is` controls this conversion. Its value is either a vector of logicals (values are recycled if necessary), or a vector of numeric or character indices which specify which columns should not be converted to factors. Note: to suppress all conversions including those of numeric columns, set `colClasses = "character"`.
`na.strings`	a vector of strings which are to be interpreted as `NA` values. Blank fields are also considered to be missing values.
`colClasses`	character. A vector of classes to be assumed for the columns. Recycled as necessary. If this is not one of the atomic vector classes (logical, integer, numeric, complex and character), there needs to be an as

	method for conversion from "character" to the specified class, or NA when type.convert is used. NB: as is in package **methods**.
nrows	the maximum number of rows to read in. Negative values are ignored.
skip	the number of lines of the data file to skip before beginning to read data.
check.names	logical. If TRUE then the names of the variables in the data frame are checked to ensure that they are syntactically valid variable names. If necessary they are adjusted (by make.names) so that they are, and also to ensure that there are no duplicates.
fill	logical. If TRUE then in case the rows have unequal length, blank fields are implicitly added. See Details.
strip.white	logical. Used only when sep has been specified, and allows the stripping of leading and trailing white space from character fields (numeric fields are always stripped). See scan for further details, remembering that the columns may include the row names.
blank.lines.skip	logical: if TRUE blank lines in the input are ignored.
comment.char	character: a character vector of length one containing a single character or an empty string. Use "" to turn off the interpretation of comments altogether.
...	Further arguments to read.table.

Details

If row.names is not specified and the header line has one less entry than the number of columns, the first column is taken to be the row names. This allows data frames to be read in from the format in which they are printed. If row.names is specified and does not refer to the first column, that column is discarded from such files.

The number of data columns is determined by looking at the first five lines of input (or the whole file if it has less than five lines), or from the length of col.names if it is specified and is longer. This could conceivably be wrong if fill or blank.lines.skip are true.

read.csv and read.csv2 are identical to read.table except for the defaults. They are intended for reading "comma separated value" files ('.csv') or the variant used in countries that use a comma as decimal point and a semicolon as field separator. Similarly, read.delim and `read.delim2` are for reading delimited files, defaulting to the TAB character for the delimiter. Notice that `header = TRUE` and `fill = TRUE` in these variants.

Comment characters are allowed unless `comment.char = ""`, and complete comment lines are allowed provided `blank.lines.skip = TRUE` However, comment lines prior to the header must have the comment character in the first non-blank column.

Value

A data frame (`data.frame`) containing a representation of the data in the file. Empty input is an error unless `col.names` is specified, when a 0-row data frame is returned: similarly giving just a header line if `header = TRUE` results in a 0-row data frame.

This function is the principal means of reading tabular data into R.

Note

The columns referred to in `as.is` and `colClasses` include the column of row names (if any).

Less memory will be used if `colClasses` is specified as one of the five atomic vector classes.

Using `nrows`, even as a mild over-estimate, will help memory usage.

Using `comment.char = ""` will be appreciably faster.

`read.table` is not the right tool for reading large matrices, especially those with many columns: it is designed to read *data frames* which may have columns of very different classes. Use `scan` instead.

References

Chambers, J. M. (1992) *Data for models.* Chapter 3 of *Statistical Models in S* eds J. M. Chambers and T. J. Hastie, Wadsworth & Brooks/Cole.

See Also

The *R Data Import/Export* manual.

`scan`, `type.convert`, `read.fwf` for reading *f*ixed *w*idth *f*ormatted input; `write.table`; `data.frame`.

`count.fields` can be useful to determine problems with reading files which result in reports of incorrect record lengths.

readBin *Transfer Binary Data To and From Connections*

Description

Read binary data from a connection, or write binary data to a connection.

Usage

```
readBin(con, what, n = 1, size = NA, signed = TRUE,
        endian = .Platform$endian)
writeBin(object, con, size = NA, endian = .Platform$endian)

readChar(con, nchars)
writeChar(object, con, nchars = nchar(object), eos = "")
```

Arguments

`con`	A connection object or a character string.
`what`	Either an object whose mode will give the mode of the vector to be read, or a character vector of length one describing the mode: one of `"numeric"`, `"double"`, `"integer"`, `"int"`, `"logical"`, `"complex"`, `"character"`.
`n`	integer. The (maximal) number of records to be read. You can use an over-estimate here, but not too large as storage is reserved for `n` items.
`size`	integer. The number of bytes per element in the byte stream. The default, `NA`, uses the natural size. Size changing is not supported for complex vectors.
`signed`	logical. Only used for integers of sizes 1 and 2, when it determines if the quantity on file should be regarded as a signed or unsigned integer.
`endian`	The endian-ness (`"big"` or `"little"` of the target system for the file. Using `"swap"` will force swapping endian-ness.
`object`	An R object to be written to the connection.
`nchars`	integer, giving the lengths of (unterminated) character strings to be read or written.

`eos`	character. The terminator to be written after each string, followed by an ASCII `nul`; use `NULL` for no terminator at all.

Details

If the `con` is a character string, the functions call `file` to obtain a file connection which is opened for the duration of the function call.

If the connection is open it is read/written from its current position. If it is not open, it is opened for the duration of the call and then closed again.

If `size` is specified and not the natural size of the object, each element of the vector is coerced to an appropriate type before being written or as it is read. Possible sizes are 1, 2, 4 and possibly 8 for integer or logical vectors, and 4, 8 and possibly 12/16 for numeric vectors. (Note that coercion occurs as signed types except if `signed = FALSE` when reading integers of sizes 1 and 2.) Changing sizes is unlikely to preserve `NA`s, and the extended precision sizes are unlikely to be portable across platforms.

`readBin` and `writeBin` read and write C-style zero-terminated character strings. Input strings are limited to 10000 characters. `readChar` and `writeChar` allow more flexibility, and can also be used on text-mode connections.

Handling R's missing and special (`Inf`, `-Inf` and `NaN`) values is discussed in the *R Data Import/Export* manual.

Value

For `readBin`, a vector of appropriate mode and length the number of items read (which might be less than `n`).

For `readChar`, a character vector of length equal to the number of items read (which might be less than `length(nchars)`).

For `writeBin` and `writeChar`, none.

Note

Integer read/writes of size 8 will be available if either C type `long` is of size 8 bytes or C type `long long` exists and is of size 8 bytes.

Real read/writes of size `sizeof(long double)` (usually 12 or 16 bytes) will be available only if that type is available and different from `double`.

Note that as R character strings cannot contain ASCII `nul`, strings read by `readChar` which contain such characters will appear to be shorter than requested, but the additional bytes are read from the file.

If the character length requested for `readChar` is longer than the string, as from version 1.4.0 what is available is returned.

If `readBin(what=character())` is used incorrectly on a file which does not contain C-style character strings, warnings (usually many) are given as from version 1.6.2. The input will be broken into pieces of length 10000 with any final part being discarded.

See Also

The *R Data Import/Export* manual.

`connections`, `readLines`, `writeLines`.

`.Machine` for the sizes of `long`, `long long` and `long double`.

Examples

```
zz <- file("testbin", "wb")
writeBin(1:10, zz)
writeBin(pi, zz, endian="swap")
writeBin(pi, zz, size=4)
writeBin(pi^2, zz, size=4, endian="swap")
writeBin(pi+3i, zz)
writeBin("A test of a connection", zz)
z <- paste("A very long string", 1:100, collapse=" + ")
writeBin(z, zz)
if(.Machine$sizeof.long==8 || .Machine$sizeof.longlong==8)
    writeBin(as.integer(5^(1:10)), zz, size = 8)
if((s <-.Machine$sizeof.longdouble) > 8)
   writeBin((pi/3)^(1:10), zz, size = s)
close(zz)

zz <- file("testbin", "rb")
readBin(zz, integer(), 4)
readBin(zz, integer(), 6)
readBin(zz, numeric(), 1, endian="swap")
readBin(zz, numeric(), size=4)
readBin(zz, numeric(), size=4, endian="swap")
readBin(zz, complex(), 1)
readBin(zz, character(), 1)
z2 <- readBin(zz, character(), 1)
```

```
if(.Machine$sizeof.long==8 || .Machine$sizeof.longlong==8)
    readBin(zz, integer(), 10,  size = 8)
if((s <-.Machine$sizeof.longdouble) > 8)
   readBin(zz, numeric(), 10, size = s)
close(zz)
unlink("testbin")
stopifnot(z2 == z)

## test fixed-length strings
zz <- file("testbin", "wb")
x <- c("a", "this will be truncated", "abc")
nc <- c(3, 10, 3)
writeChar(x, zz, nc, eos=NULL)
writeChar(x, zz, eos="\r\n")
close(zz)

zz <- file("testbin", "rb")
readChar(zz, nc)
# need to read the terminator explicitly
readChar(zz, nchar(x)+3)
close(zz)
unlink("testbin")

## signed vs unsigned ints
zz <- file("testbin", "wb")
x <- as.integer(seq(0, 255, 32))
writeBin(x, zz, size=1)
writeBin(x, zz, size=1)
x <- as.integer(seq(0, 60000, 10000))
writeBin(x, zz, size=2)
writeBin(x, zz, size=2)
close(zz)
zz <- file("testbin", "rb")
readBin(zz, integer(), 8, size=1)
readBin(zz, integer(), 8, size=1, signed=FALSE)
readBin(zz, integer(), 7, size=2)
readBin(zz, integer(), 7, size=2, signed=FALSE)
close(zz)
unlink("testbin")
```

readline *Read a Line from the Terminal*

Description

readline reads a line from the terminal

Usage

```
readline(prompt = "")
```

Arguments

prompt	the string printed when prompting the user for input. Should usually end with a space " ".

Details

The prompt string will be truncated to a maximum allowed length, normally 256 chars (but can be changed in the source code).

Value

A character vector of length one.

Examples

```
fun <- function() {
  ANSWER <- readline("Are you a satisfied R user? ")
  if (substr(ANSWER, 1, 1) == "n")
    cat("This is impossible.  YOU LIED!\n")
  else
    cat("I knew it.\n")
}
fun()
```

`readLines` *Read Text Lines from a Connection*

Description

Read text lines from a connection.

Usage

```
readLines(con = stdin(), n = -1, ok = TRUE)
```

Arguments

`con`	A connection object or a character string.
`n`	integer. The (maximal) number of lines to read. Negative values indicate that one should read up to the end of the connection.
`ok`	logical. Is it OK to reach the end of the connection before `n > 0` lines are read? If not, an error will be generated.

Details

If the `con` is a character string, the functions call `file` to obtain a file connection which is opened for the duration of the function call.

If the connection is open it is read from its current position. If it is not open, it is opened for the duration of the call and then closed again.

If the final line is incomplete (no final EOL marker) the behaviour depends on whether the connection is blocking or not. For a blocking text-mode connection (or a non-text-mode connection) the line will be accepted, with a warning. For a non-blocking text-mode connection the incomplete line is pushed back, silently.

Value

A character vector of length the number of lines read.

See Also

`connections`, `writeLines`, `readBin`, `scan`

Examples

```
cat("TITLE extra line", "2 3 5 7", "", "11 13 17",
    file="ex.data", sep="\n")
readLines("ex.data", n=-1)
unlink("ex.data") # tidy up

## difference in blocking
cat("123\nabc", file = "test1")
readLines("test1") # line with a warning

con <- file("test1", "r", blocking = FALSE)
readLines(con) # empty
cat(" def\n", file = "test1", append = TRUE)
readLines(con) # gets both
close(con)

unlink("test1") # tidy up
```

`real` *Real Vectors*

Description

`real` creates a double precision vector of the specified length. Each element of the vector is equal to `0`.

`as.real` attempts to coerce its argument to be of real type.

`is.real` returns `TRUE` or `FALSE` depending on whether its argument is of real type or not.

Usage

```
real(length = 0)
as.real(x, ...)
is.real(x)
```

Arguments

`length`	desired length.
`x`	object to be coerced or tested.
`...`	further arguments passed to or from other methods.

Note

R *has no single precision data type. All real numbers are stored in double precision format.*

Recall *Recursive Calling*

Description

Recall is used as a placeholder for the name of the function in which it is called. It allows the definition of recursive functions which still work after being renamed, see example below.

Usage

```
Recall(...)
```

Arguments

...	all the arguments to be passed.

See Also

do.call and call.

Examples

```
## A trivial (but inefficient!) example:
fib <- function(n)
  if(n<=2) {if(n>=0) 1 else 0}
  else Recall(n-1) + Recall(n-2)
fibonacci <- fib; rm(fib)
## renaming wouldn't work without Recall
fibonacci(10) # 55
```

`recover` *Browsing after an Error*

Description

This function allows the user to browse directly on any of the currently active function calls, and is suitable as an error option. The expression `options(error=recover)` will make this the error option.

Usage

```
recover()
```

Details

When called, `recover` prints the list of current calls, and prompts the user to select one of them. The standard R `browser` is then invoked from the corresponding environment; the user can type ordinary S language expressions to be evaluated in that environment.

When finished browsing in this call, type `c` to return to `recover` from the browser. Type another frame number to browse some more, or type `0` to exit `recover`.

The use of `recover` largely supersedes `dump.frames` as an error option, unless you really want to wait to look at the error. If `recover` is called in non-interactive mode, it behaves like `dump.frames`. For computations involving large amounts of data, `recover` has the advantage that it does not need to copy out all the environments in order to browse in them. If you do decide to quit interactive debugging, call `dump.frames` directly while browsing in any frame (see the examples).

WARNING: The special `Q` command to go directly from the browser to the prompt level of the evaluator currently interacts with `recover` to effectively turn off the error option for the next error (on subsequent errors, `recover` will be called normally).

Value

Nothing useful is returned. However, you *can* invoke `recover` directly from a function, rather than through the error option shown in the examples. In this case, execution continues after you type `0` to exit `recover`.

Compatibility Note

The R `recover` function can be used in the same way as the S-Plus function of the same name; therefore, the error option shown is a compatible way to specify the error action. However, the actual functions are essentially unrelated and interact quite differently with the user. The navigating commands `up` and `down` do not exist in the R version; instead, exit the browser and select another frame.

References

John M. Chambers (1998). *Programming with Data*; Springer.
See the compatibility note above, however.

See Also

`browser` for details about the interactive computations; `options` for setting the error option; `dump.frames` to save the current environments for later debugging.

Examples

```
options(error = recover) # setting the error option

### Example of interaction
> myFit <- lm(y ~ x, data = xy, weights = w)
Error in lm.wfit(x, y, w, offset = offset, ...) :
        missing or negative weights not allowed

Enter a frame number, or 0 to exit
1:lm(y ~ x, data = xy, weights = w)
2:lm.wfit(x, y, w, offset = offset, ...)
Selection: 2
Called from: eval(expr, envir, enclos)
Browse[1]> objects() # all the objects in this frame
[1] "method" "n"      "ny"     "offset" "tol"    "w"
[7] "x"      "y"
Browse[1]> w
[1] -0.5013844  1.3112515  0.2939348 -0.8983705 -0.1538642
[6] -0.9772989  0.7888790 -0.1919154 -0.3026882
Browse[1]> dump.frames() # save for offline debugging
Browse[1]> c # exit the browser

Enter a frame number, or 0 to exit
```

```
1:lm(y ~ x, data = xy, weights = w)
2:lm.wfit(x, y, w, offset = offset, ...)
Selection: 0 # exit recover
>
```

`reg.finalizer` *Finalization of objects*

Description

Registers an R function to be called upon garbage collection of object.

Usage

```
reg.finalizer(e, f)
```

Arguments

`e`	Object to finalize. Must be environment or external pointer.
`f`	Function to call on finalization. Must accept a single argument, which will be the object to finalize.

Value

`NULL`.

Note

The purpose of this function is mainly to allow objects that refer to external items (a temporary file, say) to perform cleanup actions when they are no longer referenced from within R. This only makes sense for objects that are never copied on assignment, hence the restriction to environments and external pointers.

Examples

```
f <- function(e) print("cleaning....")
g <- function(x){e<-environment(); reg.finalizer(e,f)}
g()
invisible(gc()) # trigger cleanup
```

regex *Regular Expressions as used in R*

Description

This help page documents the regular expression patterns supported by grep and related functions regexpr, sub and gsub, as well as by strsplit.

This is preliminary documentation.

Details

A 'regular expression' is a pattern that describes a set of strings. Three types of regular expressions are used in R, *extended* regular expressions, used by grep(extended = TRUE) (its default), *basic* regular expressions, as used by grep(extended = FALSE), and *Perl-like* regular expressions used by grep(perl = TRUE).

Other functions which use regular expressions (often via the use of grep) include apropos, browseEnv, help.search, list.files, ls and strsplit. These will all use *extended* regular expressions, unless strsplit is called with argument extended = FALSE.

Patterns are described here as they would be printed by cat: do remember that backslashes need to be doubled in entering R character strings from the keyboard.

Extended Regular Expressions

This section covers the regular expressions allowed if extended = TRUE in grep, regexpr, sub, gsub and strsplit. They use the GNU implementation of the POSIX 1003.2 standard.

Regular expressions are constructed analogously to arithmetic expressions, by using various operators to combine smaller expressions.

The fundamental building blocks are the regular expressions that match a single character. Most characters, including all letters and digits, are regular expressions that match themselves. Any metacharacter with special meaning may be quoted by preceding it with a backslash. The metacharacters are . \ | ( ) [{ ^ $ * + ?.

A *character class* is a list of characters enclosed by [and] matches any single character in that list; if the first character of the list is the caret ^, then it matches any character *not* in the list. For example, the regular expression [0123456789] matches any single digit, and [^abc] matches anything except the characters a, b or c. A range of characters may be specified by giving the first and last characters, separated by a hyphen. (Character ranges are interpreted in the collation order of the current locale.)

Certain named classes of characters are predefined. Their interpretation depends on the *locale* (see locales); the interpretation below is that of the POSIX locale.

`[:alnum:]` Alphanumeric characters: `[:alpha:]` and `[:digit:]`.

`[:alpha:]` Alphabetic characters: `[:lower:]` and `[:upper:]`.

`[:blank:]` Blank characters: space and tab.

`[:cntrl:]` Control characters. In ASCII, these characters have octal codes 000 through 037, and 177 (DEL). In another character set, these are the equivalent characters, if any.

`[:digit:]` Digits: `0 1 2 3 4 5 6 7 8 9`.

`[:graph:]` Graphical characters: `[:alnum:]` and `[:punct:]`.

`[:lower:]` Lower-case letters in the current locale.

`[:print:]` Printable characters: `[:alnum:]`, `[:punct:]` and space.

`[:punct:]` Punctuation characters: ``! " # $ % & ' ( ) * + , - . / : ; < = > ? @ [ \ ] ^ _ ` { | } ~``.

`[:space:]` Space characters: tab, newline, vertical tab, form feed, carriage return, and space.

`[:upper:]` Upper-case letters in the current locale.

`[:xdigit:]` Hexadecimal digits: `0 1 2 3 4 5 6 7 8 9 A B C D E F a b c d e f`.

For example, `[[:alnum:]]` means `[0-9A-Za-z]`, except the latter depends upon the locale and the character encoding, whereas the former is independent of locale and character set. (Note that the brackets in these class names are part of the symbolic names, and must be included in addition to the brackets delimiting the bracket list.) Most metacharacters lose their special meaning inside lists. To include a literal `]`, place it first in the list. Similarly, to include a literal `^`, place it anywhere but first. Finally, to include a literal `-`, place it first or last. (Only these and \ remain special inside character classes.)

The period `.` matches any single character. The symbol `\w` is documented to be synonym for `[[:alnum:]]` and `\W` is its negation. However, `\w` also matches underscore in the GNU grep code used in R.

The caret `^` and the dollar sign `$` are metacharacters that respectively match the empty string at the beginning and end of a line. The symbols `\<` and `\>` respectively match the empty string at the beginning and end of a word. The symbol `\b` matches the empty string at the edge of a word, and `\B` matches the empty string provided it is not at the edge of a word.

A regular expression may be followed by one of several repetition quantifiers:

`?` The preceding item is optional and will be matched at most once.

`*` The preceding item will be matched zero or more times.

`+` The preceding item will be matched one or more times.

`{n}` The preceding item is matched exactly `n` times.

`{n,}` The preceding item is matched `n` or more times.

`{n,m}` The preceding item is matched at least `n` times, but not more than `m` times.

Repetition is greedy, so the maximal possible number of repeats is used.

Two regular expressions may be concatenated; the resulting regular expression matches any string formed by concatenating two substrings that respectively match the concatenated subexpressions.

Two regular expressions may be joined by the infix operator `|`; the resulting regular expression matches any string matching either subexpression. For example, `abba|cde` matches either the string `abba` or the string `cde`. Note that alternation does not work inside character classes, where `|` has its literal meaning.

Repetition takes precedence over concatenation, which in turn takes precedence over alternation. A whole subexpression may be enclosed in parentheses to override these precedence rules.

The backreference `\N`, where N is a single digit, matches the substring previously matched by the Nth parenthesized subexpression of the regular expression.

The current code attempts to support traditional usage by assuming that { is not special if it would be the start of an invalid interval specification. (POSIX allows this behaviour as an extension but we advise users not to rely on it.)

Basic Regular Expressions

This section covers the regular expressions allowed if `extended = FALSE` in `grep`, `regexpr`, `sub`, `gsub` and `strsplit`.

In basic regular expressions the metacharacters ?, +, {, |, (, and) lose their special meaning; instead use the backslashed versions \?, \+, \{, \|, \(, and \). Thus the metacharacters are `. \ [ ^ $ *`.

Perl Regular Expressions

The `perl = TRUE` argument to **grep**, **regexpr**, **sub** and **gsub** switches to the PCRE library that implements regular expression pattern matching using the same syntax and semantics as Perl 5, with just a few differences. Character tables created in the C locale at compile time are used in this version, but locale-specific tables will be used in later versions of R.

For complete details please consult the man pages for PCRE (especially `man pcrepattern` or if that does not exist, `man pcre`) on your system or from the sources at `ftp://ftp.csx.cam.ac.uk/pub/software/programming/pcre/`. If PCRE support was compiled from the sources within R, the PCRE version is 3.9 as described here: PCRE $\geq$ 4.0 supports more of the Perl regular expressions.

All the regular expressions described for extended regular expressions are accepted except \< and \>: in Perl all backslashed metacharacters are alphanumeric and backslashed symbols always are interpreted as a literal character. { is not special if it would be the start of an invalid interval specification. There can be more than 9 backreferences.

The construct `(?...)` is used for Perl extensions in a variety of ways depending on what immediately follows the ?.

Perl-like matching can work in several modes, set by the options `(?i)` (caseless, equivalent to Perl's `/i`), `(?m)` (multiline, equivalent to Perl's `/m`), `(?s)` (single line, so a dot matches all characters, even new lines: equivalent to Perl's `/s`) and `(?x)` (extended, whitespace data characters are ignored unless escaped and comments are allowed: equivalent to Perl's `/x`). These can be concatenated, so for example, `(?im)` sets caseless multiline matching. It is also possible to unset these options by preceding the letter with a hyphen, and to combine setting and unsetting such as `(?im-sx)`. These settings can be applied within patterns, and then apply to the remainder of the pattern. Additional options not in Perl include `(?U)` to set 'ungreedy' mode (so matching is minimal unless ? is used, when it is greedy). Initially none of these options are set.

The escape sequences `\d`, `\s` and `\w` represent any decimal digit, space character and and 'word' character (letter, digit or underscore in the current locale) respectively, and their upper-case versions represent their negation. In PCRE 3.9 the vertical tab is not regarded as a whitespace character, but it is in PCRE $\geq$ 4.0. (Perl itself changed around version 5.004.)

Escape sequence `\a` is `BEL`, `\e` is `ESC`, `\f` is `FF`, `\n` is `LF`, `\r` is `CR` and `\t` is `TAB`. In addition `\cx` is `cntrl-x` for any `x`, `\ddd` is the octal character `ddd` (for up to three digits unless interpretable as a backreference), and `\xhh` specifies a character in hex.

Outside a character class, `\b` matches a word boundary, `\B` is its negation, `\A` matches at start of subject (even in multiline mode, unlike `^`), `\Z` matches at end of a subject or before newline at end, `\z` matches at end of a subject. and `\G` matches at first matching position in a subject. `\C` matches a single byte. including a newline.

The same repetition quantifiers as extended POSIX are supported. However, if a quantifier is followed by `?`, the match is 'ungreedy', that is as short as possible rather than as long as possible (unless the meanings are reversed by the `(?U)` option.)

The sequence `(?#` marks the start of a comment which continues up to the next closing parenthesis. Nested parentheses are not permitted. The characters that make up a comment play no part at all in the pattern matching.

If the extended option is set, an unescaped `#` character outside a character class introduces a comment that continues up to the next newline character in the pattern.

The pattern `(?:...)` groups characters just as parentheses do but does not make a backreference.

Patterns `(?=...)` and `(?!...)` are zero-width positive and negative lookahead *assertions*: they match if an attempt to match the `...` forward from the current position would succeed (or not), but use up no characters in the string being processed. Patterns `(?<=...)` and `(?<!...)` are the lookbehind equivalents: they do not allow repetition quantifiers nor `\C` in `...`.

Named subpatterns, atomic grouping, possessive qualifiers and conditional and recursive patterns are not covered here.

Author(s)

This help page is based on the documentation of GNU grep 2.4.2, from which the C code used by R has been taken, the `pcre` man page from PCRE 3.9 and the `pcrepattern` man page from PCRE 4.4.

See Also

`grep`, `apropos`, `browseEnv`, `help.search`, `list.files`, `ls` and `strsplit`.

REMOVE *Remove Add-on Packages*

Description

Utility for removing add-on packages.

Usage

```
R CMD REMOVE [options] [-l lib] pkgs
```

Arguments

`pkgs`	a list with the names of the packages to be removed.
`lib`	the path name of the R library tree to remove from. May be absolute or relative.
`options`	further options.

Details

If used as `R CMD REMOVE pkgs` without explicitly specifying `lib`, packages are removed from the library tree rooted at the first directory given in `$R_LIBS` if this is set and non-null, and to the default library tree (which is rooted at '`$R_HOME/library`') otherwise.

To remove from the library tree `lib`, use `R CMD REMOVE -l lib pkgs`.

Use `R CMD REMOVE --help` for more usage information.

See Also

`INSTALL`

remove *Remove Objects from a Specified Environment*

Description

remove and rm can be used to remove objects. These can be specified successively as character strings, or in the character vector list, or through a combination of both. All objects thus specified will be removed.

If envir is NULL then the currently active environment is searched first.

If inherits is TRUE then parents of the supplied directory are searched until a variable with the given name is encountered. A warning is printed for each variable that is not found.

Usage

```
remove(..., list = character(0), pos = -1,
       envir = as.environment(pos),
       inherits = FALSE)
rm(..., list = character(0), pos = -1,
   envir = as.environment(pos),
   inherits = FALSE)
```

Arguments

...	the objects to be removed, supplied individually and/or as a character vector
list	a character vector naming objects to be removed.
pos	where to do the removal. By default, uses the current environment. See the details for other possibilities.
envir	the environment to use. See the details section.
inherits	should the enclosing frames of the environment be inspected?

Details

The pos argument can specify the environment from which to remove the objects in any of several ways: as an integer (the position in the search list); as the character string name of an element in the search list; or as an `environment` (including using `sys.frame` to access the currently active function calls). The `envir` argument is an alternative way to specify an environment, but is primarily there for back compatibility.

References

Becker, R. A., Chambers, J. M. and Wilks, A. R. (1988) *The New S Language*. Wadsworth & Brooks/Cole.

See Also

`ls`, `objects`

Examples

```
tmp <- 1:4
## work with tmp and cleanup
rm(tmp)

## remove (almost) everything in the working environment.
## You will get no warning, so don't do this unless you are
## really sure.
rm(list = ls())
```

`remove.packages` *Remove Installed Packages*

Description

Removes installed packages and updates index information as necessary.

Usage

```
remove.packages(pkgs, lib, version)
```

Arguments

`pkgs`	a character vector with the names of the packages to be removed.
`lib`	a character string giving the library directory to move the packages from.
`version`	A character string specifying a version of the package to remove. If none is provided, the system will remove an unversioned install of the package.

See Also

`REMOVE` for a command line version; `install.packages` for installing packages.

`replace` *Replace Values in a Vector*

Description

`replace` replaces the values in `x` with indexes given in `list` by those given in `values`. If necessary, the values in `values` are recycled.

Usage

```
replace(x, list, values)
```

Arguments

`x`	vector
`list`	an index vector
`values`	replacement values

Value

A vector with the values replaced.

Note

`x` is unchanged: remember to assign the result.

References

Becker, R. A., Chambers, J. M. and Wilks, A. R. (1988) *The New S Language*. Wadsworth & Brooks/Cole.

reshape *Reshape Grouped Data*

Description

This function reshapes a data frame between 'wide' format with repeated measurements in separate columns of the same record and 'long' format with the repeated measurements in separate records.

Usage

```
reshape(data, varying = NULL, v.names = NULL,
  timevar = "time", idvar = "id", ids = 1:NROW(data),
  times = seq(length = length(varying[[1]])),
  drop = NULL, direction, new.row.names = NULL,
  split = list(regexp="\\.", include=FALSE))
```

Arguments

`data`	a data frame
`varying`	names of sets of variables in the wide format that correspond to single variables in long format ('time-varying'). A list of vectors (or optionally a matrix for `direction="wide"`). See below for more details and options.
`v.names`	names of variables in the long format that correspond to multiple variables in the wide format.
`timevar`	the variable in long format that differentiates multiple records from the same group or individual.
`idvar`	the variable in long format that identifies multiple records from the same group/individual. This variable may also be present in wide format.
`ids`	the values to use for a newly created `idvar` variable in long format.
`times`	the values to use for a newly created `timevar` variable in long format.
`drop`	a vector of names of variables to drop before reshaping
`direction`	character string, either `"wide"` to reshape to wide format, or `"long"` to reshape to long format.

`new.row.names`	logical; if `TRUE` and `direction="wide"`, create new row names in long format from the values of the id and time variables.
`split`	information for guessing the `varying`, `v.names`, and `times` arguments. See below for details.

Details

The arguments to this function are described in terms of longitudinal data, as that is the application motivating the functions. A 'wide' longitudinal dataset will have one record for each individual with some time-constant variables that occupy single columns and some time-varying variables that occupy a column for each time point. In 'long' format there will be multiple records for each individual, with some variables being constant across these records and others varying across the records. A 'long' format dataset also needs a 'time' variable identifying which time point each record comes from and an 'id' variable showing which records refer to the same person.

If the data frame resulted from a previous `reshape` then the operation can be reversed by specifying just the `direction` argument. The other arguments are stored as attributes on the data frame.

If `direction="long"` and no `varying` or `v.names` arguments are supplied it is assumed that all variables except `idvar` and `timevar` are time-varying. They are all expanded into multiple variables in wide format.

If `direction="wide"` the `varying` argument can be a vector of column names or column numbers (converted to column names). The function will attempt to guess the `v.names` and `times` from these names. The default is variable names like `x.1`, `x.2`,where `split=list(regexp="\.",include=FALSE)` specifies to split at the dot and drop it from the name. To have alphabetic followed by numeric times use `split=list(regexp="[A-Za-z][0-9]",include=TRUE)`. This splits between the alphabetic and numeric parts of the name and does not drop the regular expression.

Value

The reshaped data frame with added attributes to simplify reshaping back to the original form.

See Also

`stack`, `aperm`

Examples

```
data(Indometh,package="nls")
summary(Indometh)
wide <- reshape(Indometh, v.names="conc", idvar="Subject",
                timevar="time", direction="wide")
wide

reshape(wide, direction="long")
reshape(wide, idvar="Subject",
        varying=list(names(wide)[2:12]),
        v.names="conc", direction="long")

## times need not be numeric
df <- data.frame(id=rep(1:4,rep(2,4)),
                 visit=I(rep(c("Before","After"),4)),
                 x=rnorm(4), y=runif(4))
df
reshape(df, timevar="visit", idvar="id", direction="wide")
## warns that y is really varying
reshape(df, timevar="visit", idvar="id", direction="wide",
        v.names="x")

## unbalanced 'long' data leads to NA fill in 'wide' form
df2 <- df[1:7,]
df2
reshape(df2, timevar="visit", idvar="id", direction="wide")

## Alternative regular expressions for guessing names
df3 <- data.frame(id=1:4, age=c(40,50,60,50),
                  dose1=c(1,2,1,2), dose2=c(2,1,2,1),
                  dose4=c(3,3,3,3))
reshape(df3, direction="long", varying=3:5,
        split=list(regexp="[a-z][0-9]", include=TRUE))

## an example that isn't longitudinal data
data(state)
state.x77 <- as.data.frame(state.x77)
long <- reshape(state.x77, idvar="state",
 ids=row.names(state.x77), times=names(state.x77),
 timevar="Characteristic", varying=list(names(state.x77)),
 direction="long")
```

```
reshape(long, direction="wide")

reshape(long, direction="wide",
        new.row.names=unique(long$state))
```

rev *Reverse Elements*

Description

`rev` provides a reversed version of its argument. It is generic function with a default method for vectors and one for `dendrograms`.

Note that this is no longer needed (nor efficient) for obtaining vectors sorted into descending order, since that is now rather more directly achievable by `sort(x, decreasing=TRUE)`.

Usage

```
rev(x)
## Default S3 method:
rev(x)
```

Arguments

`x`	a vector or another object for which reversion is defined.

References

Becker, R. A., Chambers, J. M. and Wilks, A. R. (1988) *The New S Language.* Wadsworth & Brooks/Cole.

See Also

`seq`, `sort`.

Examples

```
x <- c(1:5,5:3)
## sort into descending order; first more efficiently:
stopifnot(sort(x, decreasing = TRUE) == rev(sort(x)))
stopifnot(rev(1:7) == 7:1) # don't need 'rev' here
```

RHOME *R Home Directory*

Description

Returns the location of the R home directory, which is the root of the installed R tree.

Usage

```
R RHOME
```

rle *Run Length Encoding*

Description

Compute the lengths and values of runs of equal values in a vector – or the reverse operation.

Usage

```
rle(x)
inverse.rle(x, ...)
```

Arguments

`x`	a simple vector for `rle()` or an object of class `"rle"` for `inverse.rle()`.
`...`	further arguments which are ignored in R.

Value

`rle()` returns an object of class `"rle"` which is a list with components

`lengths`	an integer vector containing the length of each run.
`values`	a vector of the same length as `lengths` with the corresponding values.

`inverse.rle()` is the inverse function of `rle()`.

Examples

```
x <- rev(rep(6:10, 1:5))
rle(x)
## lengths [1:5]  5 4 3 2 1 values   [1:5] 10 9 8 7 6

z <- c(TRUE,TRUE,FALSE,FALSE,TRUE,FALSE,TRUE,TRUE,TRUE)
rle(z)
rle(as.character(z))

stopifnot(x == inverse.rle(rle(x)),
          z == inverse.rle(rle(z)))
```

`row` *Row Indexes*

Description

Returns a matrix of integers indicating their row number in the matrix.

Usage

```
row(x, as.factor = FALSE)
```

Arguments

`x`	a matrix.
`as.factor`	a logical value indicating whether the value should be returned as a factor rather than as numeric.

Value

An integer matrix with the same dimensions as `x` and whose `ij`-th element is equal to `i`.

References

Becker, R. A., Chambers, J. M. and Wilks, A. R. (1988) *The New S Language.* Wadsworth & Brooks/Cole.

See Also

`col` to get columns.

Examples

```
x <- matrix(1:12, 3, 4)
# extract the diagonal of a matrix
dx <- x[row(x) == col(x)]
dx

# create an identity 5-by-5 matrix
x <- matrix(0, nr = 5, nc = 5)
x[row(x) == col(x)] <- 1
x
```

row.names *Get and Set Row Names for Data Frames*

Description

All data frames have a row names attribute, a character vector of length the number of rows with no duplicates nor missing values.

For convenience, these are generic functions for which users can write other methods, and there are default methods for arrays. The description here is for the `data.frame` method.

Usage

```
row.names(x)
row.names(x) <- value
```

Arguments

`x`	object of class `"data.frame"`, or any other class for which a method has been defined.
`value`	a vector with the same length as the number of rows of `x`, to be coerced to character. Duplicated or missing values are not allowed.

Value

`row.names` returns a character vector.

`row.names<-` returns a data frame with the row names changed.

Note

`row.names` is similar to `rownames` for arrays, and it has a method that calls `rownames` for an array argument.

References

Chambers, J. M. (1992) *Data for models.* Chapter 3 of *Statistical Models in S* eds J. M. Chambers and T. J. Hastie, Wadsworth & Brooks/Cole.

See Also

`data.frame`, `rownames`.

`row/colnames` *Row and Column Names*

Description

Retrieve or set the row or column names of a matrix-like object.

Usage

```
rownames(x, do.NULL = TRUE, prefix = "row")
rownames(x) <- value

colnames(x, do.NULL = TRUE, prefix = "col")
colnames(x) <- value
```

Arguments

`x`	a matrix-like R object, with at least two dimensions for `colnames`.
`do.NULL`	logical. Should this create names if they are `NULL`?
`prefix`	for created names.
`value`	a valid value for that component of `dimnames(x)`. For a matrix or array this is either `NULL` or a character vector of length the appropriate dimension.

Details

The extractor functions try to do something sensible for any matrix-like object `x`. If the object has `dimnames` the first component is used as the row names, and the second component (if any) is used for the col names. For a data frame, `rownames` and `colnames` are equivalent to `row.names` and `names` respectively.

If `do.NULL` is `FALSE`, a character vector (of length `NROW(x)` or `NCOL(x)`) is returned in any case, prepending `prefix` to simple numbers, if there are no dimnames or the corresponding component of the dimnames is `NULL`.

For a data frame, `value` for `rownames` should be a character vector of unique names, and for `colnames` a character vector of unique syntactically-valid names. (Note: uniqueness and validity are not enforced.)

See Also

`dimnames`, `case.names`, `variable.names`.

Examples

```
m0 <- matrix(NA, 4, 0)
rownames(m0)

m2 <- cbind(1,1:4)
colnames(m2, do.NULL = FALSE)
colnames(m2) <- c("x","Y")
rownames(m2) <- rownames(m2, do.NULL = FALSE,
                             prefix = "Obs.")
m2
```

`rowsum` *Give row sums of a matrix or data frame, based on a grouping variable*

Description

Compute sums across rows of a matrix-like object for each level of a grouping variable. `rowsum` is generic, with methods for matrices and data frames.

Usage

```
rowsum(x, group, reorder = TRUE, ...)
```

Arguments

`x`	a matrix, data frame or vector of numeric data. Missing values are allowed.
`group`	a vector giving the grouping, with one element per row of `x`. Missing values will be treated as another group and a warning will be given
`reorder`	if `TRUE`, then the result will be in order of `sort(unique(group))`, if `FALSE`, it will be in the order that rows were encountered.
`...`	other arguments for future methods

Details

The default is to reorder the rows to agree with `tapply` as in the example below. Reordering should not add noticeably to the time except when there are very many distinct values of `group` and `x` has few columns.

The original function was written by Terry Therneau, but this is a new implementation using hashing that is much faster for large matrices.

To add all the rows of a matrix (ie, a single `group`) use `rowSums`, which should be even faster.

Value

a matrix or data frame containing the sums. There will be one row per unique value of `group`.

See Also

tapply, aggregate,rowSums

Examples

```
x <- matrix(runif(100), ncol=5)
group <- sample(1:8, 20, TRUE)
xsum <- rowsum(x, group)
## Slower versions
xsum2 <- tapply(x, list(group[row(x)], col(x)), sum)
xsum3<- aggregate(x,list(group),sum)
```

Rprof *Enable Profiling of R's Execution*

Description

Enable or disable profiling of the execution of R expressions.

Usage

```
Rprof(filename = "Rprof.out", append = FALSE,
      interval = 0.02)
```

Arguments

`filename`	The file to be used for recording the profiling results. Set to `NULL` or `""` to disable profiling.
`append`	logical: should the file be over-written or appended to?
`interval`	real: time interval between samples.

Details

Enabling profiling automatically disables any existing profiling to another or the same file.

Profiling works by writing out the call stack every `interval` seconds, to the file specified. Either the `summaryRprof` function or the Perl script `R CMD Rprof` can be used to process the output file to produce a summary of the usage; use `R CMD Rprof --help` for usage information.

Note that the timing interval cannot be too small: once the timer goes off, the information is not recorded until the next clock tick (probably every 10msecs). Thus the interval is rounded to the nearest integer number of clock ticks, and is made to be at least one clock tick (at which resolution the total time spent is liable to be underestimated).

Note

Profiling is not available on all platforms. By default, it is attempted to compile support for profiling. Configure R with '`--disable-R-profiling`' to change this.

As R profiling uses the same mechanisms as C profiling, the two cannot be used together, so do not use `Rprof` in an executable built for profiling.

See Also

The chapter on "Tidying and profiling R code" in "Writing R Extensions" (see the 'doc/manual' subdirectory of the R source tree).

`summaryRprof`

Examples

```
Rprof()
## some code to be profiled
Rprof(NULL)
## some code NOT to be profiled
Rprof(append=TRUE)
## some code to be profiled
Rprof(NULL)
...
## Now post-process the output as described in Details
```

save *Save R Objects*

Description

save writes an external representation of R objects to the specified file. The objects can be read back from the file at a later date by using the function load (or data in some cases).

save.image() is just a short-cut for "save my current environment", i.e., save(list = ls(all=TRUE), file = ".RData"). It is what also happens with q("yes").

Usage

```
save(..., list = character(0),
     file = stop("'file' must be specified"),
     ascii = FALSE, version = NULL, envir = parent.frame(),
     compress = FALSE)
save.image(file = ".RData", version = NULL, ascii = FALSE,
           compress = FALSE, safe = TRUE)

sys.load.image(name, quiet)
sys.save.image(name)
```

Arguments

...	the names of the objects to be saved.
list	A character vector containing the names of objects to be saved.
file	a connection or the name of the file where the data will be saved. Must be a file name for workspace format version 1.
ascii	if TRUE, an ASCII representation of the data is written. This is useful for transporting data between machines of different types. The default value of ascii is FALSE which leads to a more compact binary file being written.
version	the workspace format version to use. NULL specifies the current default format. The version used from R 0.99.0 to R 1.3.1 was version 1. The default format as from R 1.4.0 is version 2.

`envir`	environment to search for objects to be saved.
`compress`	logical specifying whether saving to a named file is to use compression. Ignored when `file` is a connection and for workspace format version 1.
`safe`	logical. If `TRUE`, a temporary file is used for creating the saved workspace. The temporary file is renamed to `file` if the save succeeds. This preserves an existing workspace `file` if the save fails, but at the cost of using extra disk space during the save.
`name`	name of image file to save or load.
`quiet`	logical specifying whether a message should be printed.

Details

All R platforms use the XDR representation of binary objects in binary save-d files, and these are portable across all R platforms.

Default values for `save.image` options can be modified with the `save.image.defaults` option. This mechanism is experimental and subject to change.

`sys.save.image` is a system function that is called by `q()` and its GUI analogs; `sys.load.image` is called by the startup code. These functions should not be called directly and are subject to change.

`sys.save.image` closes all connections first, to ensure that it is able to open a connection to save the image. This is appropriate when called from `q()` and allies, but reinforces the warning that it should not be called directly.

Warning

The `...` arguments only give the *names* of the objects to be saved: they are searched for in the environment given by the `envir` argument, and the actual objects given as arguments need not be those found.

See Also

`dput`, `dump`, `load`, `data`.

Examples

```
x <- runif(20)
y <- list(a = 1, b = TRUE, c = "oops")
```

```
save(x, y, file = "xy.Rdata")
save.image()
unlink("xy.Rdata")
unlink(".RData")

# set save.image defaults using option:
options(save.image.defaults=list(ascii=TRUE, safe=FALSE))
save.image()
unlink(".RData")
```

savehistory *Load or Save or Display the Commands History*

Description

Load or save or display the commands history.

Usage

```
loadhistory(file = ".Rhistory")
savehistory(file = ".Rhistory")
history(max.show = 25, reverse = FALSE)
```

Arguments

`file`	The name of the file in which to save the history, or from which to load it. The path is relative to the current working directory.
`max.show`	The maximum number of lines to show. `Inf` will give all of the currently available history.
`reverse`	logical. If true, the lines are shown in reverse order. Note: this is not useful when there are continuation lines.

Details

This works under the `readline` and GNOME interfaces, but not if `readline` is not available (for example, in batch use).

Note

If you want to save the history (almost) every session, you can put a call to `savehistory()` in `.Last`.

Examples

```
.Last <- function()
    if(interactive()) try(savehistory("~/.Rhistory"))
```

scale *Scaling and Centering of Matrix-like Objects*

Description

`scale` is generic function whose default method centers and/or scales the columns of a numeric matrix.

Usage

```
scale(x, center = TRUE, scale = TRUE)
```

Arguments

`x`	a numeric matrix(like object).
`center`	either a logical value or a numeric vector of length equal to the number of columns of `x`.
`scale`	either a logical value or a numeric vector of length equal to the number of columns of `x`.

Details

The value of `center` determines how column centering is performed. If `center` is a numeric vector with length equal to the number of columns of `x`, then each column of `x` has the corresponding value from `center` subtracted from it. If `center` is `TRUE` then centering is done by subtracting the column means (omitting `NA`s) of `x` from their corresponding columns, and if `center` is `FALSE`, no centering is done.

The value of `scale` determines how column scaling is performed (after centering). If `scale` is a numeric vector with length equal to the number of columns of `x`, then each column of `x` is divided by the corresponding value from `scale`. If `scale` is `TRUE` then scaling is done by dividing the (centered) columns of `x` by their root-mean-square, and if `scale` is `FALSE`, no scaling is done.

The root-mean-square for a column is obtained by computing the square-root of the sum-of-squares of the non-missing values in the column divided by the number of non-missing values minus one.

Value

For `scale.default`, the centered, scaled matrix. The numeric centering and scalings used (if any) are returned as attributes `"scaled:center"` and `"scaled:scale"`

References

Becker, R. A., Chambers, J. M. and Wilks, A. R. (1988) *The New S Language*. Wadsworth & Brooks/Cole.

See Also

`sweep` which allows centering (and scaling) with arbitrary statistics.

For working with the scale of a plot, see `par`.

Examples

```
x <- matrix(1:10, nc=2)
(centered.x <- scale(x, scale=FALSE))
cov(centered.scaled.x <- scale(x)) # all 1
```

scan *Read Data Values*

Description

Read data into a vector or list from the console or file.

Usage

```
scan(file = "", what = double(0), nmax = -1, n = -1,
  sep = "", quote = if (sep=="\n") "" else "'\"", dec = ".",
  skip = 0, nlines = 0, na.strings = "NA", flush = FALSE,
  fill = FALSE, strip.white = FALSE, quiet = FALSE,
  blank.lines.skip = TRUE, multi.line = TRUE,
  comment.char = "")
```

Arguments

`file` the name of a file to read data values from. If the specified file is `""`, then input is taken from the keyboard (in this case input can be terminated by a blank line or an EOF signal, `Ctrl-D` on Unix and `Ctrl-Z` on Windows.).

Otherwise, the file name is interpreted *relative* to the current working directory (given by `getwd()`), unless it specifies an *absolute* path. Tilde-expansion is performed where supported.

Alternatively, `file` can be a `connection`, which will be opened if necessary, and if so closed at the end of the function call.

`file` can also be a complete URL.

`what` the type of `what` gives the type of data to be read. If `what` is a list, it is assumed that the lines of the data file are records each containing `length(what)` items ("fields"). The supported types are `logical`, `integer`, `numeric`, `complex`, `character` and `list`: `list` values should have elements which are one of the first five types listed or `NULL`.

`nmax` the maximum number of data values to be read, or if `what` is a list, the maximum number of records to be read. If omitted (and `nlines` is not set to a positive value), `scan` will read to the end of `file`.

`n`	the maximum number of data values to be read, defaulting to no limit.
`sep`	by default, scan expects to read white-space delimited input fields. Alternatively, `sep` can be used to specify a character which delimits fields. A field is always delimited by a newline unless it is quoted.
`quote`	the set of quoting characters as a single character string.
`dec`	decimal point character.
`skip`	the number of lines of the input file to skip before beginning to read data values.
`nlines`	the maximum number of lines of data to be read.
`na.strings`	character vector. Elements of this vector are to be interpreted as missing (`NA`) values.
`flush`	logical: if `TRUE`, `scan` will flush to the end of the line after reading the last of the fields requested. This allows putting comments after the last field, but precludes putting more that one record on a line.
`fill`	logical: if `TRUE`, `scan` will implicitly add empty fields to any lines with fewer fields than implied by `what`.
`strip.white`	vector of logical value(s) corresponding to items in the `what` argument. It is used only when `sep` has been specified, and allows the stripping of leading and trailing white space from `character` fields (`numeric` fields are always stripped). If `strip.white` is of length 1, it applies to all fields; otherwise, if `strip.white[i]` is `TRUE` *and* the `i`-th field is of mode character (because `what[i]` is) then the leading and trailing white space from field `i` is stripped.
`quiet`	logical: if `FALSE` (default), scan() will print a line, saying how many items have been read.
`blank.lines.skip`	logical: if `TRUE` blank lines in the input are ignored, except when counting `skip` and `nlines`.
`multi.line`	logical. Only used if `what` is a list. If `FALSE`, all of a record must appear on one line (but more than one record can appear on a single line). Note that using `fill = TRUE` implies that a record will terminated at the end of a line.

`comment.char`	character: a character vector of length one containing a single character or an empty string. Use `""` to turn off the interpretation of comments altogether (the default).

Details

The value of `what` can be a list of types, in which case `scan` returns a list of vectors with the types given by the types of the elements in `what`. This provides a way of reading columnar data. If any of the types is `NULL`, the corresponding field is skipped (but a `NULL` component appears in the result).

The type of `what` or its components can be one of the five atomic types or `NULL`,

Empty numeric fields are always regarded as missing values. Empty character fields are scanned as empty character vectors, unless `na.strings` contains `""` when they are regarded as missing values.

If `sep` is the default (`""`), the character \ in a quoted string escapes the following character, so quotes may be included in the string by escaping them.

If `sep` is non-default, the fields may be quoted in the style of '`.csv`' files where separators inside quotes ('' or `""`) are ignored and quotes may be put inside strings by doubling them. However, if `sep = "\n"` it is assumed by default that one wants to read entire lines verbatim.

Quoting is only interpreted in character fields, and as from R 1.8.0 in `NULL` fields (which might be skipping character fields).

Note that since `sep` is a separator and not a terminator, reading a file by `scan("foo", sep="\n", blank.lines.skip=FALSE)` will give an empty file line if the file ends in a linefeed and not if it does not. This might not be what you expected; see also `readLines`.

If `comment.char` occurs (except inside a quoted character field), it signals that the rest of the line should be regarded as a comment and be discarded. Lines beginning with a comment character (possibly after white space) are treated as blank lines.

Value

if `what` is a list, a list of the same length and same names (as any) as `what`.

Otherwise, a vector of the type of `what`.

Note

The default for `multi.line` differs from S. To read one record per line, use `flush = TRUE` and `multi.line = FALSE`.

If number of items is not specified, the internal mechanism re-allocates memory in powers of two and so could use up to three times as much memory as needed. (It needs both old and new copies.) If you can, specify either `n` or `nmax` whenever inputting a large vector, and `nmax` or `nlines` when inputting a large list.

References

Becker, R. A., Chambers, J. M. and Wilks, A. R. (1988) *The New S Language*. Wadsworth & Brooks/Cole.

See Also

`read.table` for more user-friendly reading of data matrices; `readLines` to read a file a line at a time. `write`.

Examples

```
cat("TITLE extra line", "2 3 5 7", "11 13 17",
    file="ex.data", sep="\n")
pp <- scan("ex.data", skip = 1, quiet= TRUE)
    scan("ex.data", skip = 1)
# only 1 line after the skipped one
    scan("ex.data", skip = 1, nlines=1)
str(scan("ex.data", what = list("","","")))
str(scan("ex.data", what = list("","",""), flush = TRUE))
unlink("ex.data") # tidy up
```

sd *Standard Deviation*

Description

This function computes the standard deviation of the values in `x`. If `na.rm` is `TRUE` then missing values are removed before computation proceeds. If `x` is a matrix or a data frame, a vector of the standard deviation of the columns is returned.

Usage

```
sd(x, na.rm = FALSE)
```

Arguments

`x`	a numeric vector, matrix or data frame.
`na.rm`	logical. Should missing values be removed?

See Also

`var` for its square, and `mad`, the most robust alternative.

Examples

```
sd(1:2) ^ 2
```

se.aov *Internal Functions Used by model.tables*

Description

Internal function for use by `model.tables`.

Usage

```
se.aov(object, n, type = "means")
se.aovlist(object, dn.proj, dn.strata, factors, mf,
           efficiency, n, type = "diff.means", ...)
```

See Also

`model.tables`

`search` *Give Search Path for R Objects*

Description

Gives a list of `attached` *packages* (see `library`), and R objects, usually `data.frames`.

Usage

```
search()
searchpaths()
```

Value

A character vector, starting with `".GlobalEnv"`, and ending with `"package:base"` which is R's **base** package required always.

`searchpaths` gives a similar character vector, with the entries for packages being the path to the package used to load the code.

References

Becker, R. A., Chambers, J. M. and Wilks, A. R. (1988) *The New S Language.* Wadsworth & Brooks/Cole. (`search`.)

Chambers, J. M. (1998) *Programming with Data. A Guide to the S Language.* Springer. (`searchPaths`.)

See Also

`attach` and `detach` to change the search "path", `objects` to find R objects in there.

Examples

```
search()
searchpaths()
```

seek *Functions to Reposition Connections*

Description

Functions to re-position connections.

Usage

```
seek(con, ...)
## S3 method for class 'connection':
seek(con, where = NA, origin = "start", rw = "", ...)

isSeekable(con)

truncate(con, ...)
```

Arguments

`con`	a connection.
`where`	integer. A file position (relative to the origin specified by `origin`), or `NA`.
`rw`	character. Empty or `"read"` or `"write"`, partial matches allowed.
`origin`	character. One of `"start"`, `"current"`, `"end"`.
`...`	further arguments passed to or from other methods.

Details

`seek` with `where = NA` returns the current byte offset of a connection (from the beginning), and with a non-missing `where` argument the connection is re-positioned (if possible) to the specified position. `isSeekable` returns whether the connection in principle supports `seek`: currently only (possibly compressed) file connections do.

File connections can be open for both writing/appending, in which case R keeps separate positions for reading and writing. Which `seek` refers to can be set by its `rw` argument: the default is the last mode (reading or writing) which was used. Most files are only opened for reading or writing and so default to that state. If a file is open for reading and writing but has not been used, the default is to give the reading position (0).

The initial file position for reading is always at the beginning. The initial position for writing is at the beginning of the file for modes `"r+"` and `"r+b"`, otherwise at the end of the file. Some platforms only allow writing at the end of the file in the append modes.

`truncate` truncates a file opened for writing at its current position. It works only for `file` connections, and is not implemented on all platforms.

Value

`seek` returns the current position (before any move), as a byte offset, if relevant, or `0` if not.

`truncate` returns `NULL`: it stops with an error if it fails (or is not implemented).

`isSeekable` returns a logical value, whether the connection is support `seek`.

See Also

`connections`

`seq` *Sequence Generation*

Description

Generate regular sequences.

Usage

```
from:to
seq(from, to)
seq(from, to, by=)
seq(from, to, length=)
seq(along)
```

Arguments

`from`	starting value of sequence.
`to`	(maximal) end value of the sequence.
`by`	increment of the sequence.
`length`	desired length of the sequence.
`along`	take the length from the length of this argument.

Details

The interpretation of the unnamed arguments of `seq` is *not* standard, and it is recommended to always name the arguments when programming.

The operator `:` and the first `seq(.)` form generate the sequence `from`, `from+1`, ..., `to`. `seq` is a generic function.

The second form generates `from`, `from+by`, ..., `to`.

The third generates a sequence of `length` equally spaced values from `from` to `to`.

The last generates the sequence `1`, `2`, ..., `length(along)`, unless the argument is of length 1 when it is interpreted as a `length` argument.

If `from` and `to` are factors of the same length, then `from : to` returns the "cross" of the two.

Very small sequences (with `from - to` of the order of 10^{-14} times the larger of the ends) will return `from`.

Value

The result is of mode `"integer"` if `from` is (numerically equal to an) integer and `by` is not specified.

References

Becker, R. A., Chambers, J. M. and Wilks, A. R. (1988) *The New S Language.* Wadsworth & Brooks/Cole.

See Also

`rep`, `sequence`, `row`, `col`.

Examples

```
1:4
pi:6 # float
6:pi # integer

seq(0,1, length=11)
str(seq(rnorm(20)))
seq(1,9, by = 2) # match
seq(1,9, by = pi) # stay below
seq(1,6, by = 3)
seq(1.575, 5.125, by=0.05)
seq(17) # same as 1:17

for (x in list(NULL, letters[1:6], list(1,pi)))
  cat("x=", deparse(x), ";  seq(along = x):",
      seq(along = x), "\n")

f1 <- gl(2,3); f1
f2 <- gl(3,2); f2
f1:f2 # a factor, the "cross"  f1 x f2
```

---

sequence *Create A Vector of Sequences*

Description

For each element of `nvec` the sequence `seq(nvec[i])` is created. These are appended and the result returned.

Usage

```
sequence(nvec)
```

Arguments

`nvec`	an integer vector each element of which specifies the upper bound of a sequence.

See Also

`gl`, `seq`, `rep`.

Examples

```
sequence(c(3,2)) # the concatenated sequences 1:3 and 1:2.
# [1] 1 2 3 1 2
```

serialize *Simple Serialization Interface*

Description

A simple low level interface for serializing to connections.

Usage

```
serialize(object, connection, ascii = FALSE, refhook = NULL)
unserialize(connection, refhook = NULL)
.saveRDS(object, file = "", ascii = FALSE, version = NULL,
         compress = FALSE, refhook = NULL)
.readRDS(file, refhook = NULL)
```

Arguments

`object`	R object to serialize.
`file`	a connection or the name of the file where the R object is saved to or read from.
`ascii`	a logical. If `TRUE`, an ASCII representation is written; otherwise (default), a more compact binary one is used.
`version`	the workspace format version to use. `NULL` specifies the current default format. The version used from R 0.99.0 to R 1.3.1 was version 1. The default format as from R 1.4.0 is version 2.
`compress`	a logical specifying whether saving to a named file is to use compression. Ignored when `file` is a connection and for workspace format version 1.
`connection`	an open connection.
`refhook`	a hook function for handling reference objects.

Details

The function `serialize` writes `object` to the specified connection. Sharing of reference objects is preserved within the object but not across separate calls to serialize. If `connection` is `NULL` then `object` is serialized to a scaler string, which is returned as the result of `serialize`. For a text mode connection, the default value of `ascii` is set to `TRUE`.

`unserialize` reads an object from `connection`. `connection` may also be a scaler string.

The `refhook` functions can be used to customize handling of non-system reference objects (all external pointers and weak references, and all environments other than name space and package environments and `.GlobalEnv`). The hook function for `serialize` should return a character vector for references it wants to handle; otherwise it should return `NULL`. The hook for `unserialize` will be called with character vectors supplied to `serialize` and should return an appropriate object.

Warning

These functions are still experimental. Both names, interfaces and values might change in future versions. `.saveRDS` and `.readRDS` are intended for internal use.

Examples

```
x<-serialize(list(1,2,3),NULL)
unserialize(x)
```

sets *Set Operations*

Description

Performs set union, intersection, (asymmetric!) difference, equality and membership on two vectors.

Usage

```
union(x, y)
intersect(x, y)
setdiff(x, y)
setequal(x, y)
is.element(el, set)
```

Arguments

`x, y, el, set`	vectors (of the same mode) containing a sequence of items (conceptually) with no duplicated values.

Details

Each of `union`, `intersect` and `setdiff` will remove any duplicated values in the arguments.

`is.element(x, y)` is identical to `x %in% y`.

Value

A vector of the same `mode` as `x` or `y` for `setdiff` and `intersect`, respectively, and of a common mode for `union`.

A logical scalar for `setequal` and a logical of the same length as `x` for `is.element`.

See Also

`%in%`

Examples

```
(x <- c(sort(sample(1:20, 9)),NA))
(y <- c(sort(sample(3:23, 7)),NA))
union(x, y)
```

```
intersect(x, y)
setdiff(x, y)
setdiff(y, x)
setequal(x, y)

## True for all possible x & y :
setequal( union(x,y),
          c(setdiff(x,y), intersect(x,y), setdiff(y,x)))

is.element(x, y) # length 10
is.element(y, x) # length  8
```

SHLIB *Build Shared Library for Dynamic Loading*

Description

Compile given source files using `R CMD COMPILE`, and then link all specified object files into a shared library which can be loaded into R using `dyn.load` or `library.dynam`.

Usage

```
R CMD SHLIB [options] [-o libname] files
```

Arguments

`files`	a list specifying the object files to be included in the shared library. You can also include the name of source files, for which the object files are automagically made from their sources.
`libname`	the full name of the shared library to be built, including the extension (typically '`.so`' on Unix systems). If not given, the name of the library is taken from the first file.
`options`	Further options to control the processing, or for obtaining information about usage and version of the utility.

See Also

`COMPILE`, `dyn.load`, `library.dynam`

showConnections *Display Connections*

Description

Display aspects of connections.

Usage

```
showConnections(all=FALSE)
getConnection(what)
closeAllConnections()

stdin()
stdout()
stderr()
```

Arguments

`all`	logical: if true all connections, including closed ones and the standard ones are displayed. If false only open user-created connections are included.
`what`	integer: a row number of the table given by `showConnections`.

Details

`stdin()`, `stdout()` and `stderr()` are standard connections corresponding to input, output and error on the console respectively (and not necessarily to file streams). They are text-mode connections of class `"terminal"` which cannot be opened or closed, and are read-only, write-only and write-only respectively. The `stdout()` and `stderr()` connections can be re-directed by `sink`.

`showConnections` returns a matrix of information. If a connection object has been lost or forgotten, `getConnection` will take a row number from the table and return a connection object for that connection, which can be used to close the connection, for example.

`closeAllConnections` closes (and destroys) all open user connections, restoring all `sink` diversions as it does so.

Value

`stdin()`, `stdout()` and `stderr()` return connection objects.

`showConnections` returns a character matrix of information with a row for each connection, by default only for open non-standard connections.

`getConnection` returns a connection object, or `NULL`.

See Also

`connections`

Examples

```
showConnections(all = TRUE)

textConnection(letters)
# oops, I forgot to record that one
showConnections()
#  class     description      mode  text  isopen   read  write
#3 "letters" "textConnection" "r" "text" "opened" "yes" "no"
close(getConnection(3))

showConnections()
```

`Signals` *Interrupting Execution of R*

Description

On receiving `SIGUSR1` R will save the workspace and quit. `SIGUSR2` has the same result except that the `.Last` function and `on.exit` expressions will not be called.

Usage

```
kill -USR1 pid
kill -USR2 pid
```

Arguments

`pid`	The process ID of the R process

Warning

It is possible that one or more R objects will be undergoing modification at the time the signal is sent. These objects could be saved in a corrupted form.

sink *Send R Output to a File*

Description

`sink` diverts R output to a connection.

`sink.number()` reports how many diversions are in use.

`sink.number(type = "message")` reports the number of the connection currently being used for error messages.

Usage

```
sink(file = NULL, append = FALSE,
     type = c("output", "message"))
sink.number(type = c("output", "message"))
```

Arguments

`file`	a connection or a character string naming the file to write to, or `NULL` to stop sink-ing.
`append`	logical. If `TRUE`, output will be appended to `file`; otherwise, it will overwrite the contents of `file`.
`type`	character. Either the output stream or the messages stream.

Details

`sink` diverts R output to a connection. If `file` is a character string, a file connection with that name will be established for the duration of the diversion.

Normal R output is diverted by the default `type = "output"`. Only prompts and warning/error messages continue to appear on the terminal. The latter can diverted by `type = "message"` (see below).

`sink()` or `sink(file=NULL)` ends the last diversion (of the specified type). As from R version 1.3.0 there is a stack of diversions for normal output, so output reverts to the previous diversion (if there was one). The stack is of up to 21 connections (20 diversions).

If `file` is a connection if will be opened if necessary.

Sink-ing the messages stream should be done only with great care. For that stream `file` must be an already open connection, and there is no stack of connections.

Value

sink returns NULL.

For sink.number() the number (0, 1, 2, ...) of diversions of output in place.

For sink.number("message") the connection number used for messages, 2 if no diversion has been used.

Warning

Don't use a connection that is open for sink for any other purpose. The software will stop you closing one such inadvertently.

Do not sink the messages stream unless you understand the source code implementing it and hence the pitfalls.

References

Becker, R. A., Chambers, J. M. and Wilks, A. R. (1988) *The New S Language.* Wadsworth & Brooks/Cole.

Chambers, J. M. (1998) *Programming with Data. A Guide to the S Language.* Springer.

See Also

capture.output

Examples

```
sink("sink-examp.txt")
i <- 1:10
outer(i, i, "*")
sink()
unlink("sink-examp.txt")

## capture all the output to a file.
zz <- file("all.Rout", open="wt")
sink(zz)
sink(zz, type="message")
try(log("a"))
## back to the console
sink(type="message")
sink()
try(log("a"))
```

slice.index *Slice Indexes in an Array*

Description

Returns a matrix of integers indicating the number of their slice in a given array.

Usage

```
slice.index(x, MARGIN)
```

Arguments

`x`	an array. If `x` has no dimension attribute, it is considered a one-dimensional array.
`MARGIN`	an integer giving the dimension number to slice by.

Value

An integer array `y` with dimensions corresponding to those of `x` such that all elements of slice number `i` with respect to dimension `MARGIN` have value `i`.

See Also

`row` and `col` for determining row and column indexes; in fact, these are special cases of `slice.index` corresponding to `MARGIN` equal to 1 and 2, respectively.

Examples

```
x <- array(1 : 24, c(2, 3, 4))
slice.index(x, 2)
```

slotOp *Extract Slots*

Description

Extract the contents of a slot in an object with a formal class structure.

Usage

```
object@name
```

Arguments

`object`	An object from a formally defined class.
`name`	The character-string name of the slot.

Details

These operators support the formal classes of package **methods**. See `slot` for further details. Currently there is no checking that the object is an instance of a class.

See Also

`Extract`, `slot`

`socketSelect` *Wait on Socket Connections*

Description

Waits for the first of several socket connections to become available.

Usage

```
socketSelect(socklist, write = FALSE, timeout = NULL)
```

Arguments

`socklist`	list of open socket connections
`write`	logical. If `TRUE` wait for corresponding socket to become available for writing; otherwise wait for it to become available for reading.
`timeout`	numeric or `NULL`. Time in seconds to wait for a socket to become available; `NULL` means wait indefinitely.

Details

The values in `write` are recycled if necessary to make up a logical vector the same length as `socklist`. Socket connections can appear more than once in `socklist`; this can be useful if you want to determine whether a socket is available for reading or writing.

Value

Logical the same length as `socklist` indicating whether the corresponding socket connection is available for output or input, depending on the corresponding value of `write`.

Examples

```
## test whether socket connection s is available for
## writing or reading
socketSelect(list(s,s),c(TRUE,FALSE),timeout=0)
```

source *Read R Code from a File or a Connection*

Description

source causes R to accept its input from the named file (the name must be quoted). Input is read from that file until the end of the file is reached. parse is used to scan the expressions in, they are then evaluated sequentially in the chosen environment.

Usage

```
source(file, local = FALSE, echo = verbose,
       print.eval = echo, verbose = getOption("verbose"),
       prompt.echo = getOption("prompt"),
       max.deparse.length = 150, chdir = FALSE)
```

Arguments

file
: a connection or a character string giving the name of the file or URL to read from.

local
: if local is FALSE, the statements scanned are evaluated in the user's workspace (the global environment), otherwise in the environment calling source.

echo
: logical; if TRUE, each expression is printed after parsing, before evaluation.

print.eval
: logical; if TRUE, the result of eval(i) is printed for each expression i; defaults to echo.

verbose
: if TRUE, more diagnostics (than just echo = TRUE) are printed during parsing and evaluation of input, including extra info for **each** expression.

prompt.echo
: character; gives the prompt to be used if echo = TRUE.

max.deparse.length
: integer; is used only if echo is TRUE and gives the maximal length of the "echo" of a single expression.

chdir
: logical; if TRUE, the R working directory is changed to the directory containing file for evaluating.

Details

All versions of R accept input from a connection with end of line marked by LF (as used on Unix), CRLF (as used on DOS/Windows) or CR (as used on Mac). The final line can be incomplete, that is missing the final EOL marker.

If `options`("keep.source") is true (the default), the source of functions is kept so they can be listed exactly as input. This imposes a limit of 128K chars on the function size and a nesting limit of 265. Use `option(keep.source = FALSE)` when these limits might take effect: if exceeded they generate an error.

References

Becker, R. A., Chambers, J. M. and Wilks, A. R. (1988) *The New S Language*. Wadsworth & Brooks/Cole.

See Also

`demo` which uses `source`; `eval`, `parse` and `scan`; `options("keep.source")`.

split *Divide into Groups*

Description

split divides the data in the vector x into the groups defined by f. The assignment forms replace values corresponding to such a division. Unsplit reverses the effect of split.

Usage

```
split(x, f)
split(x, f) <- value
unsplit(value, f)
```

Arguments

x	vector or data frame containing values to be divided into groups.
f	a "factor" such that factor(f) defines the grouping, or a list of such factors in which case their interaction is used for the grouping.
value	a list of vectors or data frames compatible with a splitting of x

Details

split and split<- are generic functions with default and data.frame methods.

f is recycled as necessary and if the length of x is not a multiple of the length of f a warning is printed. unsplit works only with lists of vectors. The data frame method can also be used to split a matrix into a list of matrices, and the assignment form likewise, provided they are invoked explicitly.

Value

The value returned from split is a list of vectors containing the values for the groups. The components of the list are named by the factor levels given be f. If f is longer than x some of these will be of zero length. The assignment forms return their right hand side. unsplit returns a vector for which split(x, f) equals value

References

Becker, R. A., Chambers, J. M. and Wilks, A. R. (1988) *The New S Language.* Wadsworth & Brooks/Cole.

See Also

`cut`

Examples

```
n <- 10; nn <- 100
g <- factor(round(n * runif(n * nn)))
x <- rnorm(n * nn) + sqrt(as.numeric(g))
xg <- split(x, g)
boxplot(xg, col = "lavender", notch = TRUE, varwidth = TRUE)
sapply(xg, length)
sapply(xg, mean)

## Calculate z-scores by group
z <- unsplit(lapply(split(x, g), scale), g)
tapply(z, g, mean)

# or
z <- x
split(z, g) <- lapply(split(x, g), scale)
tapply(z, g, sd)

## Split a matrix into a list by columns
ma <- cbind(x = 1:10, y = (-4:5)^2)
split(ma, col(ma))

split(1:10, 1:2)
```

sprintf *Use C-style String Formatting Commands*

Description

A wrapper for the C function `sprintf`, that returns a character vector of length one containing a formatted combination of text and variable values.

Usage

```
sprintf(fmt, ...)
```

Arguments

`fmt`	a format string.
`...`	values to be passed into `fmt`. Only logical, integer, real and character vectors are accepted, and only the first value is read from each vector.

Details

This is a wrapper for the system `sprintf` C-library function. Attempts are made to check that the mode of the values passed match the format supplied, and R's special values (`NA`, `Inf`, `-Inf` and `NaN`) are handled correctly.

The following is abstracted from K&R (see References, below). The string `fmt` contains normal characters, which are passed through to the output string, and also special characters that operate on the arguments provided through `...`. Special characters start with a `%` and terminate with one of the letters in the set `difeEgGs%`. These letters denote the following types:

`d,i` Integer value

`f` Double precision value, in decimal notation of the form "[-]mmm.ddd". The number of decimal places is specified by the precision: the default is 6; a precision of 0 suppresses the decimal point.

`e,E` Double precision value, in decimal notation of the form `[-]m.ddde[+-]xx` or `[-]m.dddE[+-]xx` `g,G` Double precision value, in `%e` or `%E` format if the exponent is less than -4 or greater than or equal to the precision, and `%f` format otherwise

`s` Character string

`%` Literal `%` (none of the formatting characters given below are permitted in this case)

In addition, between the initial `%` and the terminating conversion character there may be, in any order:

`m.n` Two numbers separated by a period, denoting the field width (`m`) and the precision (`n`)

`-` Left adjustment of converted argument in its field

`+` Always print number with sign

a space Prefix a space if the first number is not a sign

`0` For numbers, pad to the field width with leading zeros

Value

A character vector of length one. Character `NA`s are converted to `"NA"`.

Author(s)

Original code by Jonathan Rougier

References

Kernighan, B. W. and Ritchie, D. M. (1988) *The C Programming Language.* Second edition, Prentice Hall. describes the format options in table B-1 in the Appendix.

See Also

`formatC` for a way of formatting vectors of numbers in a similar fashion.

`paste` for another way of creating a vector combining text and values.

Examples

```
## be careful with the format: most things in R are floats
sprintf("%s is %f feet tall\n", "Sven", 7) # OK
try(sprintf("%s is %i feet tall\n", "Sven", 7)) # not OK
sprintf("%s is %i feet tall\n", "Sven", as.integer(7)) # OK
```

```
## use a literal % :
sprintf("%.0f%% said yes (out of a sample of size %.0f)",
        66.666, 3)

## various formats of pi :
sprintf("%f", pi)
sprintf("%.3f", pi)
sprintf("%1.0f", pi)
sprintf("%5.1f", pi)
sprintf("%05.1f", pi)
sprintf("%+f", pi)
sprintf("% f", pi)
sprintf("%-10f", pi) # left justified
sprintf("%e", pi)
sprintf("%E", pi)
sprintf("%g", pi)
sprintf("%g",   1e6 * pi) # -> exponential
sprintf("%.9g", 1e6 * pi) # -> "fixed"
sprintf("%G", 1e-6 * pi)

## no truncation:
sprintf("%1.f",101)

## More sophisticated:
lapply(c("a", "ABC", "and an even longer one"),
       function(ch) sprintf("10-string '%10s'", ch))

sapply(1:18, function(n)
 sprintf(paste("e with %2d digits = %.",n,"g",sep=""),
              n, exp(1)))
```

`sQuote` *Quote Text*

Description

Single or double quote text by combining with appropriate single or double left and right quotation marks.

Usage

```
sQuote(x)
dQuote(x)
```

Arguments

`x`	an R object, to be coerced to a character vector.

Details

The purpose of the functions is to provide a simple means of markup for quoting text to be used in the R output, e.g., in warnings or error messages.

The choice of the appropriate quotation marks depends on both the locale and the available character sets. Older Unix/X11 fonts displayed the grave accent (0x60) and the apostrophe (0x27) in a way that they could also be used as matching open and close single quotation marks. Using modern fonts, or non-Unix systems, these characters no longer produce matching glyphs. Unicode provides left and right single quotation mark characters (U+2018 and U+2019); if Unicode cannot be assumed, it seems reasonable to use the apostrophe as an undirectional single quotation mark.

Similarly, Unicode has left and right double quotation mark characters (U+201C and U+201D); if only ASCII's typewriter characteristics can be employed, than the ASCII quotation mark (0x22) should be used as both the left and right double quotation mark.

`sQuote` and `dQuote` currently only provide undirectional ASCII quotation style, but may be enhanced in the future.

References

Markus Kuhn, "ASCII and Unicode quotation marks". `http://www.cl.cam.ac.uk/~mgk25/ucs/quotes.html`

Examples

```
paste("argument", sQuote("x"), "must be non-zero")
```

stack *Stack or Unstack Vectors from a Data Frame or List*

Description

Stacking vectors concatenates multiple vectors into a single vector along with a factor indicating where each observation originated. Unstacking reverses this operation.

Usage

```
stack(x, ...)
## Default S3 method:
stack(x, ...)
## S3 method for class 'data.frame':
stack(x, select, ...)

unstack(x, ...)
## Default S3 method:
unstack(x, form, ...)
## S3 method for class 'data.frame':
unstack(x, form = formula(x), ...)
```

Arguments

`x`	object to be stacked or unstacked
`select`	expression, indicating variables to select from a data frame
`form`	a two-sided formula whose left side evaluates to the vector to be unstacked and whose right side evaluates to the indicator of the groups to create. Defaults to `formula(x)` in `unstack.data.frame`.
`...`	further arguments passed to or from other methods.

Details

The `stack` function is used to transform data available as separate columns in a data frame or list into a single column that can be used in an analysis of variance model or other linear model. The `unstack` function reverses this operation.

Value

`unstack` produces a list of columns according to the formula `form`. If all the columns have the same length, the resulting list is coerced to a data frame.

`stack` produces a data frame with two columns

`values`	the result of concatenating the selected vectors in `x`
`ind`	a factor indicating from which vector in `x` the observation originated

Author(s)

Douglas Bates

See Also

`lm`, `reshape`

Examples

```
data(PlantGrowth)
formula(PlantGrowth)         # check the default formula
pg <- unstack(PlantGrowth) # unstack using this formula
pg
stack(pg)                    # now put it back together
stack(pg, select = -ctrl)  # omitting one vector
```

`standardGeneric` *Formal Method System Placeholders*

Description

Routines which are primitives used with the **methods** package. They should not be used without it and do not need to be called directly in any case.

Usage

```
standardGeneric(f)
```

Details

`standardGeneric` dispatches the method defined for a generic function `f`, using the actual arguments in the frame from which it is called.

Author(s)

John Chambers

Startup *Initialization at Start of an R Session*

Description

In R, the startup mechanism is as follows.

Unless '`--no-environ`' was given on the command line, R searches for user and site files to process for setting environment variables. The name of the site file is the one pointed to by the environment variable `R_ENVIRON`; if this is unset or empty, '`$R_HOME/etc/Renviron.site`' is used (if it exists, which it does not in a "factory-fresh" installation). The user files searched for are '`.Renviron`' in the current or in the user's home directory (in that order). See **Details** for how the files are read.

Then R searches for the site-wide startup profile unless the command line option '`--no-site-file`' was given. The name of this file is taken from the value of the `R_PROFILE` environment variable. If this variable is unset, the default is '`$R_HOME/etc/Rprofile.site`', which is used if it exists (which it does not in a "factory-fresh" installation). This code is loaded into package **base**. Users need to be careful not to unintentionally overwrite objects in base, and it is normally advisable to use `local` if code needs to be executed: see the examples.

Then, unless '`--no-init-file`' was given, R searches for a file called '`.Rprofile`' in the current directory or in the user's home directory (in that order) and sources it into the user workspace.

It then loads a saved image of the user workspace from '`.RData`' if there is one (unless '`--no-restore-data`' was specified, or '`--no-restore`', on the command line).

Next, if a function `.First` is found on the search path, it is executed as `.First()`. Finally, function `.First.sys()` in the **base** package is run. This calls `require` to attach the default packages specified by `options("defaultPackages")`.

A function `.First` (and `.Last`) can be defined in appropriate '`.Rprofile`' or '`Rprofile.site`' files or have been saved in '`.RData`'. If you want a different set of packages than the default ones when you start, insert a call to `options` in the '`.Rprofile`' or '`Rprofile.site`' file. For example, `options(defaultPackages = character())` will attach no extra packages on startup. Alternatively, set `R_DEFAULT_PACKAGES=NULL` as an environment variable before running R. Using `options(defaultPackages = "")` or `R_DEFAULT_PACKAGES=""` enforces the R *system* default.

The commands history is read from the file specified by the environment variable `R_HISTFILE` (default '`.Rhistory`') unless '`--no-restore-history`' was specified (or '`--no-restore`').

The command-line flag '`--vanilla`' implies '`--no-site-file`', '`--no-init-file`', '`--no-restore`' and '`--no-environ`'.

Usage

```
.First <- function() { ...... }

.Rprofile <startup file>
```

Details

Note that there are two sorts of files used in startup: *environment files* which contain lists of environment variables to be set, and *profile files* which contain R code.

Lines in a site or user environment file should be either comment lines starting with `#`, or lines of the form `name=value`. The latter sets the environmental variable `name` to `value`, overriding an existing value. If `value` is of the form `${foo-bar}`, the value is that of the environmental variable `foo` if that exists and is set to a non-empty value, otherwise `bar`. This construction can be nested, so `bar` can be of the same form (as in `${foo-${bar-blah}}`).

Leading and trailing white space in `value` are stripped. `value` is processed in a similar way to a Unix shell. In particular quotes are stripped, and backslashes are removed except inside quotes.

Historical notes

Prior to R version 1.4.0, the environment files searched were '`.Renviron`' in the current directory, the file pointed to by `R_ENVIRON` if set, and '`.Renviron`' in the user's home directory.

Prior to R version 1.2.1, '`.Rprofile`' was sourced after '`.RData`' was loaded, although the documented order was as here.

The format for site and user environment files was changed in version 1.2.0. Older files are quite likely to work but may generate warnings on startup if they contained unnecessary `export` statements.

Values in environment files were not processed prior to version 1.4.0.

Note

The file '$R_HOME/etc/Renviron' is always read very early in the start-up processing. It contains environment variables set by R in the configure process. Values in that file can be overriden in site or user environment files: do not change '$R_HOME/etc/Renviron' itself.

See Also

.Last for final actions before termination.

For profiling code, see Rprof.

Examples

```
# Example ~/.Renviron on Unix
R_LIBS=~/R/library
PAGER=/usr/local/bin/less

# Example .Renviron on Windows
R_LIBS=C:/R/library
MY_TCLTK=yes
TCL_LIBRARY=c:/packages/Tcl/lib/tcl8.4

# Example of .Rprofile
options(width=65, digits=5)
options(show.signif.stars=FALSE)
ps.options(horizontal=FALSE)
set.seed(1234)
.First <- function() cat("\n   Welcome to R!\n\n")
.Last <- function()  cat("\n   Goodbye!\n\n")

# Example of Rprofile.site
local({
  old <- getOption("defaultPackages")
  options(defaultPackages = c(old, "MASS"))
})

## if .Renviron contains
FOOBAR="coo\bar"doh\ex"abc\"def'"

## then we get
> cat(Sys.getenv("FOOBAR"), "\n")
coo\bardoh\exabc"def'
```

stem *Stem-and-Leaf Plots*

Description

`stem` produces a stem-and-leaf plot of the values in `x`. The parameter `scale` can be used to expand the scale of the plot. A value of `scale=2` will cause the plot to be roughly twice as long as the default.

Usage

```
stem(x, scale = 1, width = 80, atom = 1e-08)
```

Arguments

`x`	a numeric vector.
`scale`	This controls the plot length.
`width`	The desired width of plot.
`atom`	a tolerance.

References

Becker, R. A., Chambers, J. M. and Wilks, A. R. (1988) *The New S Language*. Wadsworth & Brooks/Cole.

Examples

```
data(islands)
stem(islands)
stem(log10(islands))
```

stop *Stop Function Execution*

Description

`stop` stops execution of the current expression and executes an error action.

`geterrmessage` gives the last error message.

Usage

```
stop(..., call. = TRUE)
geterrmessage()
```

Arguments

`...`	character vectors (which are pasted together with no separator), a condition object, or `NULL`.
`call.`	logical, indicating if the call should become part of the error message.

Details

The error action is controlled by error handlers established within the executing code and by the current default error handler set by `options(error=)`. The error is first signaled as if using `signalCondition()`. If there are no handlers or if all handlers return, then the error message is printed (if `options("show.error.messages")` is true) and the default error handler is used. The default behaviour (the `NULL` error-handler) in interactive use is to return to the top level prompt or the top level browser, and in non-interactive use to (effectively) call `q("no", status=1, runLast=FALSE)`. The default handler stores the error message in a buffer; it can be retrieved by `geterrmessage()`. It also stores a trace of the call stack that can be retrieved by `traceback()`.

Errors will be truncated to `getOption("warning.length")` characters, default 1000.

Value

`geterrmessage` gives the last error message, as character string ending in `"\n"`.

References

Becker, R. A., Chambers, J. M. and Wilks, A. R. (1988) *The New S Language*. Wadsworth & Brooks/Cole.

See Also

`warning`, `try` to catch errors and retry, and `options` for setting error handlers. `stopifnot` for validity testing. `tryCatch` and `withCallingHandlers` can be used to establish custom handlers while executing an expression.

Examples

```
# don't stop on stop(.)  << Use with CARE! >>
options(error = expression(NULL))

iter <- 12
if(iter > 10) stop("too many iterations")

tst1 <- function(...) stop("dummy error")
tst1(1:10,long,calling,expression)

tst2 <- function(...) stop("dummy error", call. = FALSE)
tst2(1:10,long,calling,expression,but.not.seen.in.Error)

options(error = NULL) # revert to default
```

stopifnot *Ensure the 'Truth' of R Expressions*

Description

If any of the expressions in ... are not `all` `TRUE`, `stop` is called, producing an error message indicating the *first* element of ... which was not true.

Usage

```
stopifnot(...)
```

Arguments

...	any number of (`logical`) R expressions which should evaluate to `TRUE`.

Details

`stopifnot(A, B)` is conceptually equivalent to { `if(!all(A)) stop(...) ; if(!all(B)) stop(...)` }.

Value

(`NULL` if all statements in ... are `TRUE`.)

See Also

`stop`, `warning`.

Examples

```
# all TRUE
stopifnot(1 == 1, all.equal(pi, 3.14159265), 1 < 2)

m <- matrix(c(1,3,3,1), 2,2)
# all(.) |=>  TRUE
stopifnot(m == t(m), diag(m) == rep(1,2))

# "disable stop(.)"  << Use with CARE! >>
options(error = expression(NULL))

stopifnot(
```

```
 all.equal(pi, 3.141593),  2 < 2, all(1:10 < 12), "a" < "b")
stopifnot(
 all.equal(pi, 3.1415927), 2 < 2, all(1:10 < 12), "a" < "b")

options(error = NULL) # revert to default error handler
```

str *Compactly Display the Structure of an Arbitrary R Object*

Description

Compactly display the internal **str**ucture of an R object, a "diagnostic" function and an alternative to summary (and to some extent, dput). Ideally, only one line for each "basic" structure is displayed. It is especially well suited to compactly display the (abbreviated) contents of (possibly nested) lists. The idea is to give reasonable output for **any** R object. It calls args for (non-primitive) function objects.

`ls.str` and `lsf.str` are useful "versions" of `ls`, calling `str` on each object. They are not foolproof and should rather not be used for programming, but are provided for their usefulness.

Usage

```
str(object, ...)

## S3 method for class 'data.frame':
str(object, ...)

## Default S3 method:
str(object, max.level = 0, vec.len = 4, digits.d = 3,
    nchar.max = 128, give.attr = TRUE, give.length = TRUE,
    wid = getOption("width"), nest.lev = 0,
    indent.str = paste(rep(" ", max(0, nest.lev + 1)),
    collapse = ".."),
    ...)

 ls.str(pos = 1, pattern, ...,  envir = as.environment(pos),
      mode = "any", max.level = 1, give.attr = FALSE)
lsf.str(pos = 1, ..., envir = as.environment(pos))
```

Arguments

`object` any R object about which you want to have some information.

`max.level` maximal level of nesting which is applied for displaying nested structures, e.g., a list containing sub lists. Default 0: Display all nesting levels.

`vec.len`	numeric (>= 0) indicating how many "first few" elements are displayed of each vector. The number is multiplied by different factors (from .5 to 3) depending on the kind of vector. Default 4.
`digits.d`	number of digits for numerical components (as for `print`).
`nchar.max`	maximal number of characters to show for `character` strings. Longer strings are truncated, see `longch` example below.
`give.attr`	logical; if `TRUE` (default), show attributes as sub structures.
`give.length`	logical; if `TRUE` (default), indicate length (as `[1:... ]`).
`wid`	the page width to be used. The default is the currently active `options("width")`.
`nest.lev`	current nesting level in the recursive calls to `str`.
`indent.str`	the indentation string to use.
`...`	potential further arguments (required for Method/Generic reasons).
`pos`	integer indicating `search` path position.
`envir`	environment to use, see `ls`.
`pattern`	a regular expression passed to `ls`. Only names matching `pattern` are considered.
`mode`	character specifying the `mode` of objects to consider. Passed to `exists` and `get`.

Value

`str` does not return anything, for efficiency reasons. The obvious side effect is output to the terminal.

`ls.str` and `lsf.str` invisibly return a character vector of the matching names, similarly to `ls`.

Author(s)

Martin Maechler since 1990.

See Also

`summary`, `args`.

Examples

```
## The following examples show some of 'str' capabilities
str(1:12)
str(ls)
str(args) # more useful than args(args) !
data(freeny); str(freeny)
str(str)
str(.Machine, digits = 20)
str( lsfit(1:9,1:9))
str( lsfit(1:9,1:9),  max =1)
# save first; otherwise internal options() is used.
op <- options(); str(op)
need.dev <- !exists(".Device") || is.null(.Device)
if(need.dev) postscript()
str(par()); if(need.dev) graphics.off()

ch <- letters[1:12]; is.na(ch) <- 3:5
str(ch) # character NA's

nchar(longch <- paste(rep(letters,100), collapse=""))
str(longch)
str(longch, nchar.max = 52)

# how do the functions look like which I am using?
lsf.str()
# what are the structured objects I have defined?
ls.str(mode = "list")
# which base functions have "file" in their name ?
lsf.str(pos = length(search()), pattern = "file")
```

strsplit *Split the Elements of a Character Vector*

Description

Split the elements of a character vector x into substrings according to the presence of substring split within them.

Usage

```
strsplit(x, split, extended = TRUE)
```

Arguments

x	character vector, to be split.
split	character vector containing a regular expression to use as "split". If empty matches occur, in particular if split has length 0, x is split into single characters. If split has length greater than 1, it is re-cycled along x.
extended	if TRUE, extended regular expression matching is used, and if FALSE basic regular expressions are used.

Value

A list of length length(x) the i-th element of which contains the vector of splits of x[i].

See Also

paste for the reverse, grep and sub for string search and manipulation; further nchar, substr.

regular expression for the details of the pattern specification.

Examples

```
noquote(
 strsplit("A text I want to display with spaces", NULL)[[1]]
)

x <- c(as = "asfef", qu = "qwerty", "yuiop[", "b",
       "stuff.blah.yech")
```

```
# split x on the letter e
strsplit(x,"e")

unlist(strsplit("a.b.c", "."))
## [1] "" "" "" "" "" Note that 'split' is a regexp! If you
## really want to split on '.', use
unlist(strsplit("a.b.c", "\\."))
## [1] "a" "b" "c"

## a useful function: rev() for strings
strReverse <- function(x)
  sapply(lapply(strsplit(x,NULL), rev), paste, collapse="")
strReverse(c("abc", "Statistics"))

a <- readLines(file.path(R.home(),"AUTHORS"))[-(1:8)]
a <- a[(0:2)-length(a)]
sub("\t.*","", a)
strReverse(sub(" .*","", a))
```

structure *Attribute Specification*

Description

`structure` returns the given object with its attributes set.

Usage

```
structure(.Data, ...)
```

Arguments

`.Data`	an object which will have various attributes attached to it.
`...`	attributes, specified in `tag=value` form, which will be attached to data.

References

Becker, R. A., Chambers, J. M. and Wilks, A. R. (1988) *The New S Language.* Wadsworth & Brooks/Cole.

Examples

```
structure(1:6, dim = 2:3)
```

strwrap *Wrap Character Strings to Format Paragraphs*

Description

Each character string in the input is first split into paragraphs (on lines containing whitespace only). The paragraphs are then formatted by breaking lines at word boundaries. The target columns for wrapping lines and the indentation of the first and all subsequent lines of a paragraph can be controlled independently.

Usage

```
strwrap(x, width = 0.9 * getOption("width"), indent = 0,
        exdent = 0, prefix = "", simplify = TRUE)
```

Arguments

`x`	a character vector
`width`	a positive integer giving the target column for wrapping lines in the output.
`indent`	a non-negative integer giving the indentation of the first line in a paragraph.
`exdent`	a non-negative integer specifying the indentation of subsequent lines in paragraphs.
`prefix`	a character string to be used as prefix for each line.
`simplify`	a logical. If `TRUE`, the result is a single character vector of line text; otherwise, it is a list of the same length as `x` the elements of which are character vectors of line text obtained from the corresponding element of `x`. (Hence, the result in the former case is obtained by unlisting that of the latter.)

Details

Whitespace in the input is destroyed. Double spaces after periods (thought as representing sentence ends) are preserved. Currently, it possible sentence ends at line breaks are not considered specially.

Indentation is relative to the number of characters in the prefix string.

Examples

```
## Read in file 'THANKS'.
x <- paste(readLines(file.path(R.home(), "THANKS")),
           collapse = "\n")
## Split into paragraphs and remove the first three ones
x <- unlist(strsplit(x, "\n[ \t\n]*\n"))[-(1:3)]
## Join the rest
x <- paste(x, collapse = "\n\n")
## Now for some fun:
writeLines(strwrap(x, width = 60))
writeLines(strwrap(x, width = 60, indent = 5))
writeLines(strwrap(x, width = 60, exdent = 5))
writeLines(strwrap(x, prefix = "THANKS> "))
```

subset *Subsetting Vectors and Data Frames*

Description

Return subsets of vectors or data frames which meet conditions.

Usage

```
subset(x, ...)

## Default S3 method:
subset(x, subset, ...)

## S3 method for class 'data.frame':
subset(x, subset, select, ...)
```

Arguments

`x`	object to be subsetted.
`subset`	logical expression.
`select`	expression, indicating columns to select from a data frame.
`...`	further arguments to be passed to or from other methods.

Details

For ordinary vectors, the result is simply `x[subset & !is.na(subset)]`.

For data frames, the `subset` argument works similarly on the rows. Note that `subset` will be evaluated in the data frame, so columns can be referred to (by name) as variables.

The `select` argument exists only for the method for data frames. It works by first replacing names in the selection expression with the corresponding column numbers in the data frame and then using the resulting integer vector to index the columns. This allows the use of the standard indexing conventions so that for example ranges of columns can be specified easily.

Value

An object similar to `x` containing just the selected elements (for a vector), rows and columns (for a data frame), and so on.

Author(s)

Peter Dalgaard

See Also

`[`, `transform`

Examples

```
data(airquality)
subset(airquality, Temp > 80, select = c(Ozone, Temp))
subset(airquality, Day == 1, select = -Temp)
subset(airquality, select = Ozone:Wind)

with(airquality, subset(Ozone, Temp > 80))
```

substitute *Substituting and Quoting Expressions*

Description

substitute returns the parse tree for the (unevaluated) expression expr, substituting any variables bound in env.

quote simply returns its argument. The argument is not evaluated and can be any R expression.

Usage

```
substitute(expr, env=<<see below>>)
quote(expr)
```

Arguments

expr	Any syntactically valid R expression
env	An environment or a list object. Defaults to the current evaluation environment.

Details

The typical use of substitute is to create informative labels for data sets and plots. The myplot example below shows a simple use of this facility. It uses the functions deparse and substitute to create labels for a plot which are character string versions of the actual arguments to the function myplot.

Substitution takes place by examining each component of the parse tree as follows: If it is not a bound symbol in env, it is unchanged. If it is a promise object, i.e., a formal argument to a function or explicitly created using delay(), the expression slot of the promise replaces the symbol. If it is an ordinary variable, its value is substituted, unless env is .GlobalEnv in which case the symbol is left unchanged.

Value

The mode of the result is generally "call" but may in principle be any type. In particular, single-variable expressions have mode "name" and constants have the appropriate base mode.

Note

Substitute works on a purely lexical basis. There is no guarantee that the resulting expression makes any sense.

Substituting and quoting often causes confusion when the argument is `expression(...)`. The result is a call to the `expression` constructor function and needs to be evaluated with `eval` to give the actual expression object.

References

Becker, R. A., Chambers, J. M. and Wilks, A. R. (1988) *The New S Language.* Wadsworth & Brooks/Cole.

See Also

`missing` for argument "missingness", `bquote` for partial substitution, `sQuote` and `dQuote` for adding quotation marks to strings.

Examples

```
# expression(1 + b)
(s.e <- substitute(expression(a + b), list(a = 1)))
# 1 + b
(s.s <- substitute( a + b,            list(a = 1)))
c(mode(s.e), typeof(s.e)) # "call", "language"
c(mode(s.s), typeof(s.s)) #  (the same)
# but:
(e.s.e <- eval(s.e))          # expression(1 + b)
c(mode(e.s.e), typeof(e.s.e)) # "expression", "expression"

substitute(x <- x + 1, list(x=1)) # nonsense

myplot <- function(x, y)
    plot(x, y, xlab=deparse(substitute(x)),
         ylab=deparse(substitute(y)))

## Simple examples about lazy evaluation, etc:
f1 <- function(x, y = x)             { x <- x + 1; y }
s1 <- function(x, y = substitute(x)) { x <- x + 1; y }
s2 <- function(x, y) {
  if(missing(y)) y <- substitute(x); x <- x + 1; y
}
a <- 10
```

```
f1(a) # 11
s1(a) # 11
s2(a) # a
typeof(s2(a)) # "symbol"
```

`substr` *Substrings of a Character Vector*

Description

Extract or replace substrings in a character vector.

Usage

```
substr(x, start, stop)
substring(text, first, last = 1000000)
substr(x, start, stop) <- value
substring(text, first, last = 1000000) <- value
```

Arguments

`x, text`	a character vector
`start, first`	integer. The first element to be replaced.
`stop, last`	integer. The last element to be replaced.
`value`	a character vector, recycled if necessary.

Details

`substring` is compatible with S, with `first` and `last` instead of `start` and `stop`. For vector arguments, it expands the arguments cyclically to the length of the longest.

When extracting, if `start` is larger than the string length then `""` is returned.

For the replacement functions, if `start` is larger than the string length then no replacement is done. If the portion to be replaced is longer than the replacement string, then only the portion the length of the string is replaced.

Value

For `substr`, a character vector of the same length as `x`.

For `substring`, a character vector of length the longest of the arguments.

Note

The S4 version of `substring<-` ignores `last`; this version does not.

References

Becker, R. A., Chambers, J. M. and Wilks, A. R. (1988) *The New S Language.* Wadsworth & Brooks/Cole. (substring.)

See Also

strsplit, paste, nchar.

Examples

```
substr("abcdef",2,4)
substring("abcdef",1:6,1:6)
## strsplit is more efficient ...

substr(rep("abcdef",4),1:4,4:5)
x <- c("asfef", "qwerty", "yuiop[", "b", "stuff.blah.yech")
substr(x, 2, 5)
substring(x, 2, 4:6)

substring(x, 2) <- c("..", "+++")
x
```

`summary` *Object Summaries*

Description

`summary` is a generic function used to produce result summaries of the results of various model fitting functions. The function invokes particular `methods` which depend on the `class` of the first argument.

Usage

```
summary(object, ...)

## Default S3 method:
summary(object, ..., digits = max(3, getOption("digits")-3))
## S3 method for class 'data.frame':
summary(object, maxsum = 7,
       digits = max(3, getOption("digits")-3), ...)
## S3 method for class 'factor':
summary(object, maxsum = 100, ...)
## S3 method for class 'matrix':
summary(object, ...)
```

Arguments

`object`	an object for which a summary is desired.
`maxsum`	integer, indicating how many levels should be shown for `factor`s.
`digits`	integer, used for number formatting with `signif()` (for `summary.default`) or `format()` (for `summary.data.frame`).
`...`	additional arguments affecting the summary produced.

Details

For `factor`s, the frequency of the first `maxsum - 1` most frequent levels is shown, where the less frequent levels are summarized in `"(Others)"` (resulting in `maxsum` frequencies).

The functions `summary.lm` and `summary.glm` are examples of particular methods which summarise the results produced by `lm` and `glm`.

Value

The form of the value returned by **summary** depends on the class of its argument. See the documentation of the particular methods for details of what is produced by that method.

References

Chambers, J. M. and Hastie, T. J. (1992) *Statistical Models in S.* Wadsworth & Brooks/Cole.

See Also

`anova`, `summary.glm`, `summary.lm`.

Examples

```
data(attenu)
# summary.data.frame(...), default precision
summary(attenu, digits = 4)
# summary.factor(...)
summary(attenu $ station, maxsum = 20)

lst <- unclass(attenu$station) > 20 # logical with NAs
## summary.default() for logicals -- different from
## *.factor:
summary(lst)
summary(as.factor(lst))
```

`summaryRprof` *Summarise Output of R Profiler*

Description

Summarise the output of the `Rprof` function to show the amount of time used by different R functions.

Usage

```
summaryRprof(filename = "Rprof.out", chunksize = 5000)
```

Arguments

`filename`	Name of a file produced by `Rprof()`
`chunksize`	Number of lines to read at a time

Details

This function is an alternative to `R CMD Rprof`. It provides the convenience of an all-R implementation but will be slower for large files.

As the profiling output file could be larger than available memory, it is read in blocks of `chunksize` lines. Increasing `chunksize` will make the function run faster if sufficient memory is available.

Value

A list with components

`by.self`	Timings sorted by 'self' time
`by.total`	Timings sorted by 'total' time
`sampling.time`	Total length of profiling run

See Also

The chapter on "Tidying and profiling R code" in "Writing R Extensions" (see the '`doc/manual`' subdirectory of the R source tree).

`Rprof`

Examples

```
## Rprof() is not available on all platforms
Rprof(tmp <- tempfile())
example(glm)
Rprof()
summaryRprof(tmp)
unlink(tmp)
```

sweep *Sweep out Array Summaries*

Description

Return an array obtained from an input array by sweeping out a summary statistic.

Usage

```
sweep(x, MARGIN, STATS, FUN="-", ...)
```

Arguments

x	an array.
MARGIN	a vector of indices giving the extents of x which correspond to STATS.
STATS	the summary statistic which is to be swept out.
FUN	the function to be used to carry out the sweep. In the case of binary operators such as "/" etc., the function name must be quoted.
...	optional arguments to FUN.

Value

An array with the same shape as x, but with the summary statistics swept out.

References

Becker, R. A., Chambers, J. M. and Wilks, A. R. (1988) *The New S Language*. Wadsworth & Brooks/Cole.

See Also

apply on which sweep is based; scale for centering and scaling.

Examples

```
data(attitude)
med.att <- apply(attitude, 2, median)
# subtract the column medians
sweep(data.matrix(attitude), 2, med.att)
```

`switch` *Select One of a List of Alternatives*

Description

`switch` evaluates `EXPR` and accordingly chooses one of the further arguments (in `...`).

Usage

```
switch(EXPR, ...)
```

Arguments

`EXPR`	an expression evaluating to a number or a character string.
`...`	the list of alternatives, given explicitly.

Details

If the value of `EXPR` is an integer between 1 and `nargs()-1` then the corresponding element of `...` is evaluated and the result returned.

If `EXPR` returns a character string then that string is used to match the names of the elements in `...`. If there is an exact match then that element is evaluated and returned if there is one, otherwise the next element is chosen, e.g., `switch("cc", a=1, cc=, d=2)` evaluates to 2.

In the case of no match, if there's a further argument in `switch` that one is returned, otherwise `NULL`.

Warning

Beware of partial matching: an alternative `E = foo` will match the first argument `EXPR` unless that is named. See the examples for good practice in naming the first argument.

References

Becker, R. A., Chambers, J. M. and Wilks, A. R. (1988) *The New S Language*. Wadsworth & Brooks/Cole.

Examples

```
centre <- function(x, type) {
  switch(type,
         mean = mean(x),
         median = median(x),
         trimmed = mean(x, trim = .1))
}
x <- rcauchy(10)
centre(x, "mean")
centre(x, "median")
centre(x, "trimmed")

ccc <- c("b","QQ","a","A","bb")
for(ch in ccc)
 cat(ch,":",switch(EXPR = ch, a=1,    b=2:3), "\n")
for(ch in ccc)
 cat(ch,":",
     switch(EXPR = ch, a=,A=1, b=2:3, "Otherwise: last"),
     "\n")

## Numeric EXPR don't allow an 'otherwise':
for(i in c(-1:3,9))  print(switch(i, 1,2,3,4))
```

symnum *Symbolic Number Coding*

Description

Symbolically encode a given numeric or logical vector or array.

Usage

```
symnum(x, cutpoints=c(0.3, 0.6, 0.8, 0.9, 0.95),
  symbols=c(" ", ".", ",", "+", "*", "B"),
  legend = length(symbols) >= 3,
  na = "?", eps = 1e-5, corr = missing(cutpoints),
  show.max = if(corr) "1", show.min = NULL,
  abbr.colnames = has.colnames,
  lower.triangular = corr && is.numeric(x) && is.matrix(x),
  diag.lower.tri   = corr && !is.null(show.max))
```

Arguments

`x` numeric or logical vector or array.

`cutpoints` numeric vector whose values `cutpoints[j]` $= c_j$ (*after* augmentation, see `corr` below) are used for intervals.

`symbols` character vector, one shorter than (the *augmented*, see `corr` below) `cutpoints`. `symbols[j]`$= s_j$ are used as "code" for the (half open) interval $(c_j, c_{j+1}]$.

For `logical` argument `x`, the default is `c(".","|")` (graphical 0 / 1 s).

`legend` logical indicating if a `"legend"` attribute is desired.

`na` character or logical. How `NA`s are coded. If `na == FALSE`, `NA`s are coded invisibly, *including* the `"legend"` attribute below, which otherwise mentions NA coding.

`eps` absolute precision to be used at left and right boundary.

`corr` logical. If `TRUE`, `x` contains correlations. The cutpoints are augmented by `0` and `1` and `abs(x)` is coded.

`show.max` if `TRUE`, or of mode `character`, the maximal cutpoint is coded especially.

show.min
: if TRUE, or of mode character, the minimal cutpoint is coded especially.

abbr.colnames
: logical, integer or NULL indicating how column names should be abbreviated (if they are); if NULL (or FALSE and x has no column names), the column names will all be empty, i.e., ""; otherwise if abbr.colnames is false, they are left unchanged. If TRUE or integer, existing column names will be abbreviated to abbreviate(*, minlength = abbr.colnames).

lower.triangular
: logical. If TRUE and x is a matrix, only the *lower triangular* part of the matrix is coded as non-blank.

diag.lower.tri
: logical. If lower.triangular *and* this are TRUE, the *diagonal* part of the matrix is shown.

Value

An atomic character object of class noquote and the same dimensions as x.

If legend (TRUE by default when there more than 2 classes), it has an attribute "legend" containing a legend of the returned character codes, in the form

$$c_1 s_1 c_2 s_2 \ldots s_n c_{n+1}$$

where c_j = cutpoints[j] and s_j = symbols[j].

Author(s)

Martin Maechler

See Also

as.character

Examples

```
ii <- 0:8; names(ii) <- ii
symnum(ii, cut= 2*(0:4), sym = c(".", "-", "+", "$"))
symnum(ii, cut= 2*(0:4), sym = c(".", "-", "+", "$"),
       show.max=TRUE)

symnum(1:12 %% 3 == 0) # use for logical
```

```
## Symbolic correlation matrices:
data(attitude)
symnum(cor(attitude), diag = FALSE)
symnum(cor(attitude), abbr.= NULL)
symnum(cor(attitude), abbr.= FALSE)
symnum(cor(attitude), abbr.= 2)

symnum(cor(rbind(1, rnorm(25), rnorm(25)^2)))
symnum(cor(matrix(rexp(30, 1), 5, 18))) # <<-- PATTERN ! --
# < White Noise SMALL n
symnum(cm1 <- cor(matrix(rnorm(90) ,  5, 18)))
symnum(cm1, diag=FALSE)
# < White Noise "BIG" n
symnum(cm2 <- cor(matrix(rnorm(900), 50, 18)))
symnum(cm2, lower=FALSE)

## NA's:
Cm <- cor(matrix(rnorm(60),  10, 6)); Cm[c(3,6), 2] <- NA
symnum(Cm, show.max=NULL)

## Graphical P-values (aka "significance stars"):
pval <- rev(sort(c(outer(1:6, 10^-(1:3)))))
symp <- symnum(pval, corr=FALSE,
               cutpoints = c(0,  .001,.01,.05, .1, 1),
               symbols = c("***","**","*","."," "))
noquote(cbind(P.val = format(pval), Signif= symp))
```

Syntax *Operator Syntax*

Description

Outlines R syntax and gives the precedence of operators

Details

The following unary and binary operators are defined. They are listed in precedence groups, from highest to lowest.

`[ [[`	indexing
`::`	name space/variable name separator
`$ @`	component / slot extraction
`^`	exponentiation (right to left)
`- +`	unary minus and plus
`:`	sequence operator
`%any%`	special operators
`* /`	multiply, divide
`+ -`	(binary) add, subtract
`< > <= >= == !=`	ordering and comparison
`!`	negation
`& &&`	and
`| ||`	or
`~`	as in formulae
`-> ->>`	rightwards assignment
`=`	assignment (right to left)
`<- <<-`	assignment (right to left)
`?`	help (unary and binary)

Within an expression operators of equal precedence are evaluated from left to right except where indicated.

The links in the **See Also** section covers most other aspects of the basic syntax.

Note

There are substantial precedence differences between R and S. In particular, in S ? has the same precedence as + - and & && | || have equal precedence.

References

Becker, R. A., Chambers, J. M. and Wilks, A. R. (1988) *The New S Language*. Wadsworth & Brooks/Cole.

See Also

`Arithmetic`, `Comparison`, `Control`, `Extract`, `Logic`, `Paren`

The *R Language Definition* manual.

Sys.getenv *Get Environment Variables*

Description

Sys.getenv obtains the values of the environment variables named by x.

Usage

```
Sys.getenv(x)
```

Arguments

x	a character vector, or missing

Value

A vector of the same length as x, with the variable names as its names attribute. Each element holds the value of the environment variable named by the corresponding component of x (or "" if no environment variable with that name was found).

On most platforms Sys.getenv() will return a named vector giving the values of all the environment variables.

See Also

Sys.putenv, getwd for the working directory.

Examples

```
Sys.getenv(c("R_HOME", "R_PAPERSIZE", "R_PRINTCMD", "HOST"))
```

`Sys.info` *Extract System and User Information*

Description

Reports system and user information.

Usage

```
Sys.info()
```

Details

This function is not implemented on all R platforms, and returns `NULL` when not available. Where possible it is based on POSIX system calls.

`Sys.info()` returns details of the platform R is running on, whereas `R.version` gives details of the platform R was built on: they may well be different.

Value

A character vector with fields

`sysname`	The operating system.
`release`	The OS release.
`version`	The OS version.
`nodename`	A name by which the machine is known on the network (if any).
`machine`	A concise description of the hardware.
`login`	The user's login name, or `"unknown"` if it cannot be ascertained.
`user`	The name of the real user ID, or `"unknown"` if it cannot be ascertained.

The first five fields come from the `uname(2)` system call. The login name comes from `getlogin(2)`, and the user name from `getpwuid(getuid())`

Note

The meaning of OS "release" and "version" is highly system-dependent and there is no guarantee that the node or login or user names will be what you might reasonably expect. (In particular on some GNU/Linux distributions the login name is unknown from sessions with re-directed inputs.)

See Also

`.Platform`, and `R.version`.

Examples

```
Sys.info()
## An alternative (and probably better) way to get the
## login name on Unix
Sys.getenv("LOGNAME")
```

sys.parent *Functions to Access the Function Call Stack*

Description

These functions provide access to `environments` ("frames" in S terminology) associated with functions further up the calling stack.

Usage

```
sys.call(which = 0)
sys.frame(which = 0)
sys.nframe()
sys.function(n = 0)
sys.parent(n = 1)

sys.calls()
sys.frames()
sys.parents()
sys.on.exit()
sys.status()
parent.frame(n = 1)
```

Arguments

`which`	the frame number if non-negative, the number of generations to go back if negative. (See the Details section.)
`n`	the number of frame generations to go back.

Details

`.GlobalEnv` is given number 0 in the list of frames. Each subsequent function evaluation increases the frame stack by 1 and the environment for evaluation of that function is returned by `sys.frame` with the appropriate index.

The parent of a function evaluation is the environment in which the function was called. It is not necessarily numbered one less than the frame number of the current evaluation, nor is it the environment within which the function was defined. `sys.parent` returns the number of the parent frame if `n` is 1 (the default), the grandparent if `n` is 2, and so on. `sys.frame` returns the environment associated with a given frame number.

`sys.call` and `sys.frame` both accept integer values for the argument `which`. Non-negative values of `which` are normal frame numbers whereas negative values are counted back from the frame number of the current evaluation.

`sys.nframe` returns the number of the current frame in that list. `sys.function` gives the definition of the function currently being evaluated in the frame `n` generations back.

`sys.frames` gives a list of all the active frames and `sys.parents` gives the indices of the parent frames of each of the frames.

Notice that even though the `sys.`*xxx* functions (except `sys.status`) are interpreted, their contexts are not counted nor are they reported. There is no access to them.

`sys.status()` returns a list with components `sys.calls`, `sys.parents` and `sys.frames`.

`sys.on.exit()` retrieves the expression stored for use by `on.exit` in the function currently being evaluated. (Note that this differs from S, which returns a list of expressions for the current frame and its parents.)

`parent.frame(n)` is a convenient shorthand for `sys.frame(sys.parent(n))` (implemented slightly more efficiently).

References

Becker, R. A., Chambers, J. M. and Wilks, A. R. (1988) *The New S Language.* Wadsworth & Brooks/Cole. (not `parent.frame`.)

See Also

`eval` for the usage of `sys.frame` and `parent.frame`.

Examples

```
ff <- function(x) gg(x)
gg <- function(y) sys.status()
str(ff(1))

gg <- function(y) {
    ggg <- function() {
        cat("current frame is", sys.nframe(), "\n")
        cat("parents are", sys.parents(), "\n")
```

```
        print(sys.function(0)) # ggg
        print(sys.function(2)) # gg
    }
    if(y > 0) gg(y-1) else ggg()
}
gg(3)

t1 <- function() {
  aa <- "here"
  t2 <- function() {
    ## in frame 2 here
    cat("current frame is", sys.nframe(), "\n")
    str(sys.calls()) ## list with components t1() and t2()
    cat("parents are frame nos", sys.parents(), "\n") ## 0 1
    print(ls(envir=sys.frame(-1))) ## [1] "aa" "t2"
    invisible()
  }
  t2()
}
t1()

test.sys.on.exit <- function() {
  on.exit(print(1))
  ex <- sys.on.exit()
  str(ex)
  cat("exiting...\n")
}
test.sys.on.exit()
## gives 'language print(1)', prints 1 on exit
```

Sys.putenv *Set Environment Variables*

Description

putenv sets environment variables (for other processes called from within R or future calls to **Sys.getenv** from this R process).

Usage

```
Sys.putenv(...)
```

Arguments

...	arguments in **name=value** form, with **value** coercible to a character string.

Details

Non-standard R names must be quoted: see the Examples section.

Value

A logical vector of the same length as **x**, with elements being true if setting the corresponding variable succeeded.

Note

Not all systems need support **Sys.putenv**.

See Also

Sys.getenv, **setwd** for the working directory.

Examples

```
print(Sys.putenv("R_TEST"="testit", ABC=123))
Sys.getenv("R_TEST")
```

Sys.sleep — *Suspend Execution for a Time Interval*

Description

Suspend execution of R expressions for a given number of seconds

Usage

```
Sys.sleep(time)
```

Arguments

`time`	The time interval to suspend execution for, in seconds.

Details

Using this function allows R to be given very low priority and hence not to interfere with more important foreground tasks. A typical use is to allow a process launched from R to set itself up and read its input files before R execution is resumed.

The intention is that this function suspends execution of R expressions but wakes the process up often enough to respond to GUI events, typically every 0.5 seconds.

There is no guarantee that the process will sleep for the whole of the specified interval, and it may well take slightly longer in real time to resume execution. The resolution of the time interval is system-dependent, but will normally be down to 0.02 secs or better.

Value

Invisible `NULL`.

Note

This function may not be implemented on all systems.

Examples

```
testit <- function(x)
{
    p1 <- proc.time()
    Sys.sleep(x)
```

```
    proc.time() - p1 # The cpu usage should be negligible
}
testit(3.7)
```

sys.source *Parse and Evaluate Expressions from a File*

Description

Parses expressions in the given file, and then successively evaluates them in the specified environment.

Usage

```
sys.source(file, envir = NULL, chdir = FALSE,
           keep.source = getOption("keep.source.pkgs"))
```

Arguments

`file`	a character string naming the file to be read from
`envir`	an R object specifying the environment in which the expressions are to be evaluated. May also be a list or an integer. The default value `NULL` corresponds to evaluation in the base environment. This is probably not what you want; you should typically supply an explicit `envir` argument.
`chdir`	logical; if `TRUE`, the R working directory is changed to the directory containing `file` for evaluating.
`keep.source`	logical. If `TRUE`, functions "keep their source" including comments, see `options(keep.source = *)` for more details.

Details

For large files, `keep.source = FALSE` may save quite a bit of memory. In order for the code being evaluated to use the correct environment (for example, in global assignments), source code in packages should call `topenv()`, which will return the namespace, if any, the environment set up by `sys.source`, or the global environment if a saved image is being used.

See Also

`source`, and `library` which uses `sys.source`.

system *Invoke a System Command*

Description

`system` invokes the OS command specified by `command`.

Usage

```
system(command, intern = FALSE, ignore.stderr = FALSE)
```

Arguments

`command`	the system command to be invoked, as a string.
`intern`	a logical, indicates whether to make the output of the command an R object.
`ignore.stderr`	a logical indicating whether error messages (written to '`stderr`') should be ignored.

Details

If `intern` is `TRUE` then `popen` is used to invoke the command and the output collected, line by line, into an R `character` vector which is returned as the value of `system`. Output lines of more that 8096 characters will be split.

If `intern` is `FALSE` then the C function `system` is used to invoke the command and the value returned by `system` is the exit status of this function.

`unix` is a *deprecated* alternative, available for backwards compatibility.

Value

If `intern=TRUE`, a character vector giving the output of the command, one line per character string. If the command could not be run or gives an error, an R error is generated.

If `intern=FALSE`, the return value is an error code.

See Also

`.Platform` for platform specific variables.

Examples

```
# list all files in the current directory using the -F flag
system("ls -F")

# t1 is a character vector, each one
# representing a separate line of output from who
t1 <- system("who", TRUE)

# empty since file doesn't exist
system("ls fizzlipuzzli", TRUE, TRUE)
```

system.file *Find Names of R System Files*

Description

Finds the full file names of files in packages etc.

Usage

```
system.file(..., package = "base", lib.loc = NULL)
```

Arguments

`...`	character strings, specifying subdirectory and file(s) within some package. The default, none, returns the root of the package. Wildcards are not supported.
`package`	a character string with the name of a single package. An error occurs if more than one package name is given.
`lib.loc`	a character vector with path names of R libraries, or `NULL`. The default value of `NULL` corresponds to all libraries currently known. If the default is used, the loaded packages are searched before the libraries.

Value

A character vector of positive length, containing the file names that matched `...`, or the empty string, `""`, if none matched. If matching the root of a package, there is no trailing separator.

As a special case, `system.file()` gives the root of the **base** package only.

See Also

`list.files`

Examples

```
system.file() # The root of the 'base' package
system.file(package = "lqs") # The root of package 'lqs'
system.file("INDEX")
system.file("help", "AnIndex", package = "stepfun")
```

system.time *CPU Time Used*

Description

Return CPU (and other) times that `expr` used.

Usage

```
system.time(expr)
unix.time(expr)
```

Arguments

`expr`	Valid R expression to be "timed"

Details

`system.time` calls the builtin `proc.time`, evaluates `expr`, and then calls `proc.time` once more, returning the difference between the two `proc.time` calls.

The values returned by the `proc.time` are (on Unix) those returned by the C library function `times`(3v), if available.

`unix.time` is an alias of `system.time`, for compatibility reasons.

Value

A numeric vector of length 5 containing the user cpu, system cpu, elapsed, subproc1, subproc2 times. The subproc times are the user and system cpu time used by child processes (and so are usually zero).

The resolution of the times will be system-specific; it is common for them to be recorded to of the order of 1/100 second, and elapsed time is rounded to the nearest 1/100.

Note

It is possible to compile R without support for `system.time`, when all the values will be `NA`.

See Also

`proc.time`, `time` which is for time series.

Examples

```
system.time(for(i in 1:100) mad(runif(1000)))

exT <- function(n = 1000) {
  # Purpose: Test if system.time works ok; n: loop size
  system.time(for(i in 1:n) x <- mean(rt(1000, df=4)))
}
# Try to interrupt one of the following (using
# Ctrl-C / Escape):
exT()                  # about 3 secs on a 1GHz PIII
system.time(exT())     #~ +/- same
```

t *Matrix Transpose*

Description

Given a matrix or `data.frame` `x`, `t` returns the transpose of `x`.

Usage

```
t(x)
```

Arguments

`x`	a matrix or data frame, typically.

Details

A data frame is first coerced to a matrix: see `as.matrix`. When `x` is a vector, it is treated as "column", i.e., the result is a 1-row matrix.

References

Becker, R. A., Chambers, J. M. and Wilks, A. R. (1988) *The New S Language.* Wadsworth & Brooks/Cole.

See Also

`aperm` for permuting the dimensions of arrays.

Examples

```
a <- matrix(1:30, 5,6)
ta <- t(a) # i.e., a[i, j] == ta[j, i] for all i,j :
for(j in seq(ncol(a)))
  if(! all(a[, j] == ta[j, ])) stop("wrong transpose")
```

table *Cross Tabulation and Table Creation*

Description

`table` uses the cross-classifying factors to build a contingency table of the counts at each combination of factor levels.

Usage

```
table(..., exclude = c(NA, NaN), dnn = list.names(...),
      deparse.level = 1)
as.table(x, ...)
is.table(x)

## S3 method for class 'table':
as.data.frame(x, row.names = NULL, optional = FALSE, ...)
```

Arguments

`...`	objects which can be interpreted as factors (including character strings), or a list (or data frame) whose components can be so interpreted
`exclude`	values to use in the exclude argument of `factor` when interpreting non-factor objects; if specified, levels to remove from all factors in `...`.
`dnn`	the names to be given to the dimensions in the result (the *dimnames names*).
`deparse.level`	controls how the default `dnn` is constructed. See details.
`x`	an arbitrary R object, or an object inheriting from class `"table"` for the `as.data.frame` method.
`row.names`	a character vector giving the row names for the data frame.
`optional`	a logical controlling whether row names are set. Currently not used.

Details

If the argument `dnn` is not supplied, the internal function `list.names` is called to compute the 'dimname names'. If the arguments in `...` are named, those names are used. For the remaining arguments, `deparse.level = 0` gives an empty name, `deparse.level = 1` uses the supplied argument if it is a symbol, and `deparse.level = 2` will deparse the argument.

Only when `exclude` is specified (i.e., not by default), will `table` drop levels of factor arguments potentially.

Value

`table()` returns a *contingency table*, an object of `class "table"`; see the `print` method's separate documentation.

There is a `summary` method for objects created by `table` or `xtabs`, which gives basic information and performs a chi-squared test for independence of factors (note that the function `chisq.test` in package **ctest** currently only handles 2-d tables).

`as.table` and `is.table` coerce to and test for contingency table, respectively.

The `as.data.frame` method for objects inheriting from class `"table"` can be used to convert the array-based representation of a contingency table to a data frame containing the classifying factors and the corresponding counts (the latter as component `Freq`). This is the inverse of `xtabs`.

References

Becker, R. A., Chambers, J. M. and Wilks, A. R. (1988) *The New S Language*. Wadsworth & Brooks/Cole.

See Also

Use `ftable` for printing (and more) of multidimensional tables.

Examples

```
## Simple frequency distribution
table(rpois(100,5))
data(warpbreaks)
attach(warpbreaks)
## Check the design:
table(wool, tension)
data(state)
table(state.division, state.region)
detach()
```

```
data(airquality)
# simple two-way contingency table
with(airquality, table(cut(Temp, quantile(Temp)), Month))

a <- letters[1:3]
table(a, sample(a))                    # dnn is c("a", "")
table(a, sample(a), deparse.level = 0) # dnn is c("", "")
table(a, sample(a), deparse.level = 2) # dnn is
                                       # c("a", "sample(a)")

## xtabs() <-> as.data.frame.table() :
data(UCBAdmissions) ## already a contingency table
DF <- as.data.frame(UCBAdmissions)
class(tab <- xtabs(Freq ~ ., DF)) # xtabs & table
## tab *is* "the same" as the original table:
all(tab == UCBAdmissions)
all.equal(dimnames(tab), dimnames(UCBAdmissions))

a <- rep(c(NA, 1/0:3), 10)
table(a)
table(a, exclude=NULL)
b <- factor(rep(c("A","B","C"), 10))
table(b)
table(b, exclude="B")
d <- factor(rep(c("A","B","C"), 10),
            levels=c("A","B","C","D","E"))
table(d, exclude="B")

## NA counting:
is.na(d) <- 3:4
d <- factor(d, exclude=NULL)
d[1:7]
table(d, exclude = NULL)
```

tapply *Apply a Function Over a "Ragged" Array*

Description

Apply a function to each cell of a ragged array, that is to each (non-empty) group of values given by a unique combination of the levels of certain factors.

Usage

```
tapply(X, INDEX, FUN = NULL, ..., simplify = TRUE)
```

Arguments

`X`	an atomic object, typically a vector.
`INDEX`	list of factors, each of same length as `X`.
`FUN`	the function to be applied. In the case of functions like `+`, `%*%`, etc., the function name must be quoted. If `FUN` is `NULL`, tapply returns a vector which can be used to subscript the multi-way array `tapply` normally produces.
`...`	optional arguments to `FUN`.
`simplify`	If `FALSE`, `tapply` always returns an array of mode `"list"`. If `TRUE` (the default), then if `FUN` always returns a scalar, `tapply` returns an array with the mode of the scalar.

Value

When `FUN` is present, `tapply` calls `FUN` for each cell that has any data in it. If `FUN` returns a single atomic value for each cell (e.g., functions `mean` or `var`) and when `simplify` is `TRUE`, `tapply` returns a multi-way array containing the values. The array has the same number of dimensions as `INDEX` has components; the number of levels in a dimension is the number of levels (`nlevels()`) in the corresponding component of `INDEX`.

Note that contrary to S, `simplify = TRUE` always returns an array, possibly 1-dimensional.

If `FUN` does not return a single atomic value, `tapply` returns an array of mode `list` whose components are the values of the individual calls to `FUN`, i.e., the result is a list with a `dim` attribute.

References

Becker, R. A., Chambers, J. M. and Wilks, A. R. (1988) *The New S Language*. Wadsworth & Brooks/Cole.

See Also

the convenience functions `by` and `aggregate` (using `tapply`); `apply`, `lapply` with its versions `sapply` and `mapply`.

Examples

```
groups <- as.factor(rbinom(32, n = 5, p = .4))
tapply(groups, groups, length) # is almost the same as
table(groups)

data(warpbreaks)
## contingency table from data.frame : array with named
## dimnames
tapply(warpbreaks$breaks, warpbreaks[,-1], sum)
tapply(warpbreaks$breaks, warpbreaks[, 3, drop = FALSE],
       sum)

n <- 17; fac <- factor(rep(1:3, len = n), levels = 1:5)
table(fac)
tapply(1:n, fac, sum)
tapply(1:n, fac, sum, simplify = FALSE)
tapply(1:n, fac, range)
tapply(1:n, fac, quantile)

## example of ... argument: find quarterly means
data(presidents)
tapply(presidents, cycle(presidents), mean, na.rm = TRUE)

ind <- list(c(1, 2, 2), c("A", "A", "B"))
table(ind)
tapply(1:3, ind) # the split vector
tapply(1:3, ind, sum)
```

`taskCallback` *Add or remove a top-level task callback*

Description

`addTaskCallback` registers an R function that is to be called each time a top-level task is completed.

`removeTaskCallback` un-registers a function that was registered earlier via `addTaskCallback`.

These provide low-level access to the internal/native mechanism for managing task-completion actions. One can use `taskCallbackManager` at the S-language level to manage S functions that are called at the completion of each task. This is easier and more direct.

Usage

```
addTaskCallback(f, data = NULL, name = character(0))
removeTaskCallback(id)
```

Arguments

`f`	the function that is to be invoked each time a top-level task is successfully completed. This is called with 5 or 4 arguments depending on whether `data` is specified or not, respectively. The return value should be a logical value indicating whether to keep the callback in the list of active callbacks or discard it.
`data`	if specified, this is the 5-th argument in the call to the callback function `f`.
`id`	a string or an integer identifying the element in the internal callback list to be removed. Integer indices are 1-based, i.e the first element is 1. The names of currently registered handlers is available using `getTaskCallbackNames` and is also returned in a call to `addTaskCallback`.
`name`	character: names to be used.

Details

Top-level tasks are individual expressions rather than entire lines of input. Thus an input line of the form `expression1 ; expression2` will give rise to 2 top-level tasks.

A top-level task callback is called with the expression for the top-level task, the result of the top-level task, a logical value indicating whether it was successfully completed or not (always TRUE at present), and a logical value indicating whether the result was printed or not. If the `data` argument was specified in the call to `addTaskCallback`, that value is given as the fifth argument.

The callback function should return a logical value. If the value is FALSE, the callback is removed from the task list and will not be called again by this mechanism. If the function returns TRUE, it is kept in the list and will be called on the completion of the next top-level task.

Value

`addTaskCallback` returns an integer value giving the position in the list of task callbacks that this new callback occupies. This is only the current position of the callback. It can be used to remove the entry as long as no other values are removed from earlier positions in the list first.

`removeTaskCallback` returns a logical value indicating whether the specified element was removed. This can fail (i.e., return `FALSE`) if an incorrect name or index is given that does not correspond to the name or position of an element in the list.

Note

This is an experimental feature and the interface may be changed in the future.

There is also C-level access to top-level task callbacks to allow C routines rather than R functions be used.

See Also

`getTaskCallbackNames` `taskCallbackManager` `http://developer.r-project.org/TaskHandlers.pdf`

Examples

```
times <- function(total = 3, str="Task a") {
  ctr <- 0

  function(expr, value, ok, visible) {
   ctr <<- ctr + 1
   cat(str, ctr, "\n")
```

```
    if(ctr == total) {
      cat("handler removing itself\n")
    }
    return(ctr < total)
  }
}

# add the callback that will work for
# 4 top-level tasks and then remove itself.
n <- addTaskCallback(times(4))

# now remove it, assuming it is still first in the list.
removeTaskCallback(n)

# There is no point in running this as
addTaskCallback(times(4))

sum(1:10)
sum(1:10)
sum(1:10)
sum(1:10)
sum(1:10)
```

taskCallbackManager *Create an R-level task callback manager*

Description

This provides an entirely S-language mechanism for managing callbacks or actions that are invoked at the conclusion of each top-level task. Essentially, we register a single R function from this manager with the underlying, native task-callback mechanism and this function handles invoking the other R callbacks under the control of the manager. The manager consists of a collection of functions that access shared variables to manage the list of user-level callbacks.

Usage

```
taskCallbackManager(handlers = list(),
                    registered = FALSE,
                    verbose = FALSE)
```

Arguments

`handlers` this can be a list of callbacks in which each element is a list with an element named `"f"` which is a callback function, and an optional element named `"data"` which is the 5-th argument to be supplied to the callback when it is invoked. Typically this argument is not specified, and one uses `add` to register callbacks after the manager is created.

`registered` a logical value indicating whether the `evaluate` function has already been registered with the internal task callback mechanism. This is usually `FALSE` and the first time a callback is added via the `add` function, the `evaluate` function is automatically registered. One can control when the function is registered by specifying `TRUE` for this argument and calling `addTaskCallback` manually.

`verbose` a logical value, which if `TRUE`, causes information to be printed to the console about certain activities this dispatch manager performs. This is useful for debugging callbacks and the handler itself.

Value

A list containing 6 functions:

`add` register a callback with this manager, giving the function, an optional 5-th argument, an optional name by which the callback is stored in the list, and a `register` argument which controls whether the `evaluate` function is registered with the internal C-level dispatch mechanism if necessary.

`remove` remove an element from the manager's collection of callbacks, either by name or position/index.

`evaluate` the 'real' callback function that is registered with the C-level dispatch mechanism and which invokes each of the R-level callbacks within this manager's control.

`suspend` a function to set the suspend state of the manager. If it is suspended, none of the callbacks will be invoked when a task is completed. One sets the state by specifying a logical value for the `status` argument.

`register` a function to register the `evaluate` function with the internal C-level dispatch mechanism. This is done automatically by the `add` function, but can be called manually.

`callbacks` returns the list of callbacks being maintained by this manager.

Note

This is an experimental feature and the interface may be changed in the future.

See Also

`addTaskCallback` `removeTaskCallback` `getTaskCallbackNames`
`http://developer.r-project.org/TaskHandlers.pdf`

Examples

```
# create the manager
h <- taskCallbackManager()

# add a callback
h$add(function(expr, value, ok, visible) {
```

```
    cat("In handler\n")
    return(TRUE)
  }, name = "simpleHandler")

  # look at the internal callbacks.
  getTaskCallbackNames()

  # look at the R-level callbacks
  names(h$callback())

  #
  getTaskCallbackNames()
  removeTaskCallback("R-taskCallbackManager")
```

---

taskCallbackNames *Query the names of the current internal top-level task callbacks*

Description

This provides a way to get the names (or identifiers) for the currently registered task callbacks that are invoked at the conclusion of each top-level task. These identifiers can be used to remove a callback.

Usage

```
getTaskCallbackNames()
```

Arguments

Value

A character vector giving the name for each of the registered callbacks which are invoked when a top-level task is completed successfully. Each name is the one used when registering the callbacks and returned as the in the call to `addTaskCallback`.

Note

One can use `taskCallbackManager` to manage user-level task callbacks, i.e., S-language functions, entirely within the S language and access the names more directly.

See Also

`addTaskCallback` `removeTaskCallback` `taskCallbackManager` `http://developer.r-project.org/TaskHandlers.pdf`

Examples

```
n <- addTaskCallback(function(expr, value, ok, visible) {
                       cat("In handler\n")
                       return(TRUE)
                     }, name = "simpleHandler")

getTaskCallbackNames()
```

```
# now remove it by name
removeTaskCallback("simpleHandler")

h <- taskCallbackManager()
h$add(function(expr, value, ok, visible) {
                        cat("In handler\n")
                        return(TRUE)
                      }, name = "simpleHandler")
getTaskCallbackNames()
removeTaskCallback("R-taskCallbackManager")
```

tempfile *Create Names for Temporary Files*

Description

`tempfile` returns a vector of character strings which can be used as names for temporary files.

Usage

```
tempfile(pattern = "file", tmpdir = tempdir())
tempdir()
```

Arguments

`pattern`	a non-empty character vector giving the initial part of the name.
`tmpdir`	a non-empty character vector giving the directory name

Details

If `pattern` has length greater than one then the result is of the same length giving a temporary file name for each component of `pattern`.

The names are very likely to be unique among calls to `tempfile` in an R session and across simultaneous R sessions. The filenames are guaranteed not to be currently in use.

The file name is made of the pattern, the process number in hex and a random suffix in hex. By default, the filenames will be in the directory given by `tempdir()`. This will be a subdirectory of the directory given by the environment variable `TMPDIR` if set, otherwise `"/tmp"`.

Value

For `tempfile` a character vector giving the names of possible (temporary) files. Note that no files are generated by `tempfile`.

For `tempdir`, the path of the per-session temporary directory.

References

Becker, R. A., Chambers, J. M. and Wilks, A. R. (1988) *The New S Language*. Wadsworth & Brooks/Cole.

See Also

`unlink` for deleting files.

Examples

```
tempfile(c("ab", "a b c")) # give file name with spaces in!
```

textConnection *Text Connections*

Description

Input and output text connections.

Usage

```
textConnection(object, open = "r", local = FALSE)
```

Arguments

`object`	character. A description of the connection. For an input this is an R character vector object, and for an output connection the name for the R character vector to receive the output.
`open`	character. Either `"r"` (or equivalently `""`) for an input connection or `"w"` or `"a"` for an output connection.
`local`	logical. Used only for output connections. If `TRUE`, output is assigned to a variable in the calling environment. Otherwise the global environment is used.

Details

An input text connection is opened and the character vector is copied at time the connection object is created, and `close` destroys the copy.

An output text connection is opened and creates an R character vector of the given name in the user's workspace or in the calling environment, depending on the value of the `local` argument. This object will at all times hold the completed lines of output to the connection, and `isIncomplete` will indicate if there is an incomplete final line. Closing the connection will output the final line, complete or not. (A line is complete once it has been terminated by end-of-line, represented by `"\n"` in R.)

Opening a text connection with `mode = "a"` will attempt to append to an existing character vector with the given name in the user's workspace or the calling environment. If none is found (even if an object exists of the right name but the wrong type) a new character vector wil be created, with a warning.

You cannot `seek` on a text connection, and `seek` will always return zero as the position.

Value

A connection object of class "`textConnection`" which inherits from class "`connection`".

Note

As output text connections keep the character vector up to date line-by-line, they are relatively expensive to use, and it is often better to use an anonymous `file()` connection to collect output.

On platforms where `vsnprintf` does not return the needed length of output (e.g., Windows) there is a 100,000 character limit on the length of line for output connections: longer lines will be truncated with a warning.

See Also

`connections`, `showConnections`, `pushBack`, `capture.output`.

Examples

```
zz <- textConnection(LETTERS)
readLines(zz, 2)
scan(zz, "", 4)
pushBack(c("aa", "bb"), zz)
scan(zz, "", 4)
close(zz)

zz <- textConnection("foo", "w")
writeLines(c("testit1", "testit2"), zz)
cat("testit3 ", file=zz)
isIncomplete(zz)
cat("testit4\n", file=zz)
isIncomplete(zz)
close(zz)
foo

# capture R output: use part of example from help(lm)
zz <- textConnection("foo", "w")
ctl <- c(4.17, 5.58, 5.18, 6.11, 4.5, 4.61, 5.17,
         4.53, 5.33, 5.14)
trt <- c(4.81, 4.17, 4.41, 3.59, 5.87, 3.83, 6.03,
         4.89, 4.32, 4.69)
group <- gl(2, 10, 20, labels = c("Ctl", "Trt"))
```

```
weight <- c(ctl, trt)
sink(zz)
anova(lm.D9 <- lm(weight ~ group))
cat("\nSummary of Residuals:\n\n")
summary(resid(lm.D9))
sink()
close(zz)
cat(foo, sep = "\n")
```

toString *toString Converts its Argument to a Character String*

Description

This is a helper function for `format`. It converts its argument to a string. If the argument is a vector then its elements are concatenated with a , as a separator. Most methods should honor the width argument. The minimum value for `width` is six.

Usage

```
toString(x, ...)

## Default S3 method:
toString(x, width, ...)
```

Arguments

`x`	The object to be converted.
`width`	The returned value is at most the first `width` characters.
`...`	Optional arguments for methods.

Value

A character vector of length 1 is returned.

Author(s)

Robert Gentleman

See Also

`format`

Examples

```
x <- c("a", "b", "aaaaaaaaaa")
toString(x)
toString(x, width=8)
```

trace *Interactive Tracing and Debugging of Calls to a Function or Method*

Description

A call to `trace` allows you to insert debugging code (e.g., a call to `browser` or `recover`) at chosen places in any function. A call to `untrace` cancels the tracing. Specified methods can be traced the same way, without tracing all calls to the function. Trace code can be any R expression. Tracing can be temporarily turned on or off globally by calling `tracingState`.

Usage

```
trace(what, tracer, exit, at, print, signature,
      where = topenv(parent.frame()))
untrace(what, signature = NULL,
        where = topenv(parent.frame()))

tracingState(on = NULL)
```

Arguments

`what`	The name (quoted or not) of a function to be traced or untraced. More than one name can be given in the quoted form, and the same action will be applied to each one.
`tracer`	Either a function or an unevaluated expression. The function will be called or the expression will be evaluated either at the beginning of the call, or before those steps in the call specified by the argument `at`. See the details section.
`exit`	Either a function or an unevaluated expression. The function will be called or the expression will be evaluated on exiting the function. See the details section.
`at`	optional numeric vector. If supplied, `tracer` will be called just before the corresponding step in the body of the function. See the details section.
`print`	If `TRUE` (as per default), a descriptive line is printed before any trace expression is evaluated.

`signature`	If this argument is supplied, it should be a signature for a method for function `what`. In this case, the method, and *not* the function itself, is traced.
`where`	the environment in which to look for the function to be traced; by default, the top-level environment of the call to `trace`. If you put a call to `trace` into code in a package, you may need to specify `where=.GlobalEnv` if the package containing the call has a namespace, but the function you want to trace is somewhere on the search list.
`on`	logical; a call to `tracingState` returns `TRUE` if tracing is globally turned on, `FALSE` otherwise. An argument of one or the other of those values sets the state. If the tracing state is `FALSE`, none of the trace actions will actually occur (used, for example, by debugging functions to shut off tracing during debugging).

Details

The `trace` function operates by constructing a revised version of the function (or of the method, if `signature` is supplied), and assigning the new object back where the original was found. If only the `what` argument is given, a line of trace printing is produced for each call to the function (back compatible with the earlier version of `trace`).

The object constructed by `trace` is from a class that extends `"function"` and which contains the original, untraced version. A call to `untrace` re-assigns this version.

If the argument `tracer` or `exit` is the name of a function, the tracing expression will be a call to that function, with no arguments. This is the easiest and most common case, with the functions `browser` and `recover` the likeliest candidates; the former browses in the frame of the function being traced, and the latter allows browsing in any of the currently active calls.

The `tracer` or `exit` argument can also be an unevaluated expression (such as returned by a call to `quote` or `substitute`). This expression itself is inserted in the traced function, so it will typically involve arguments or local objects in the traced function. An expression of this form is useful if you only want to interact when certain conditions apply (and in this case you probably want to supply `print=FALSE` in the call to `trace` also).

When the `at` argument is supplied, it should be a vector of integers referring to the substeps of the body of the function (this only works if the body of the function is enclosed in `{ ...}`. In this case `tracer` is *not* called on entry, but instead just before evaluating each of the steps listed in `at`. (Hint: you don't want to try to count the steps in the printed version of a function; instead, look at `as.list(body(f))` to get the numbers associated with the steps in function `f`.)

An intrinsic limitation in the `exit` argument is that it won't work if the function itself uses `on.exit`, since the existing calls will override the one supplied by `trace`.

Tracing does not nest. Any call to `trace` replaces previously traced versions of that function or method, and `untrace` always restores an untraced version. (Allowing nested tracing has too many potentials for confusion and for accidentally leaving traced versions behind.)

Tracing primitive functions (builtins and specials) from the base package works, but only by a special mechanism and not very informatively. Tracing a primitive causes the primitive to be replaced by a function with argument ... (only). You can get a bit of information out, but not much. A warning message is issued when `trace` is used on a primitive.

The practice of saving the traced version of the function back where the function came from means that tracing carries over from one session to another, *if* the traced function is saved in the session image. (In the next session, `untrace` will remove the tracing.) On the other hand, functions that were in a package, not in the global environment, are not saved in the image, so tracing expires with the session for such functions.

Tracing a method is basically just like tracing a function, with the exception that the traced version is stored by a call to `setMethod` rather than by direct assignment, and so is the untraced version after a call to `untrace`.

The version of `trace` described here is largely compatible with the version in S-Plus, although the two work by entirely different mechanisms. The S-Plus `trace` uses the session frame, with the result that tracing never carries over from one session to another (R does not have a session frame). Another relevant distinction has nothing directly to do with `trace`: The browser in S-Plus allows changes to be made to the frame being browsed, and the changes will persist after exiting the browser. The R browser allows changes, but they disappear when the browser exits. This may be relevant in that the S-Plus version allows you to experiment with code changes interactively, but the R version does not. (A future revision may include a "destructive" browser for R.)

Value

The traced function(s) name(s). The relevant consequence is the assignment that takes place.

Note

The version of function tracing that includes any of the arguments except for the function name requires the methods package (because it uses special classes of objects to store and restore versions of the traced functions).

If methods dispatch is not currently on, `trace` will load the methods namespace, but will not put the methods package on the search list.

References

Becker, R. A., Chambers, J. M. and Wilks, A. R. (1988) *The New S Language*. Wadsworth & Brooks/Cole.

See Also

`browser` and `recover`, the likeliest tracing functions; also, `quote` and `substitute` for constructing general expressions.

Examples

```
f <- function(x, y) {
    y <- pmax(y, .001)
    x ^ y
}

## arrange to call the browser on entering and exiting
## function f
trace("f", browser, exit = browser)

## instead, conditionally assign some data, and then browse
## on exit, but only then.  Don't bother me otherwise
trace("f", quote(if(any(y < 0)) yOrig <- y),
      exit = quote(if(exists("yOrig")) browser()),
      print = FALSE)

## trace a utility function, with recover so we can browse
## in the calling functions as well.
trace("as.matrix", recover)
```

```
## turn off the tracing
untrace(c("f", "as.matrix"))

if(!hasMethods) detach("package:methods")
```

`traceback` *Print Call Stack of Last Error*

Description

`traceback()` prints the call stack of the last error, i.e., the sequence of calls that lead to the error. This is useful when an error occurs with an unidentifiable error message. This stack is stored as a list in `.Traceback`, which `traceback` prints in a user-friendly format.

Usage

```
traceback()
```

Value

`traceback()` returns nothing, but prints the deparsed call stack deepest call first. The calls may print on more that one line, and the first line is labelled by the frame number.

References

Becker, R. A., Chambers, J. M. and Wilks, A. R. (1988) *The New S Language.* Wadsworth & Brooks/Cole.

Examples

```
foo <- function(x) { print(1); bar(2) }
bar <- function(x) { x + a.variable.which.does.not.exist }

foo(2) # gives a strange error
traceback()
## 2: bar(2) 1: foo(2)
bar
## Ah, this is the culprit ...
```

`transform` *Transform an Object, for Example a Data Frame*

Description

`transform` is a generic function, which—at least currently—only does anything useful with data frames. `transform.default` converts its first argument to a data frame if possible and calls `transform.data.frame`.

Usage

```
transform(x, ...)
```

Arguments

`x`	The object to be transformed
`...`	Further arguments of the form `tag=value`

Details

The `...` arguments to `transform.data.frame` are tagged vector expressions, which are evaluated in the data frame `x`. The tags are matched against `names(x)`, and for those that match, the value replace the corresponding variable in `x`, and the others are appended to `x`.

Value

The modified value of `x`.

Note

If some of the values are not vectors of the appropriate length, you deserve whatever you get!

Author(s)

Peter Dalgaard

See Also

`subset`, `list`, `data.frame`

Examples

```
data(airquality)
transform(airquality, Ozone = -Ozone)
transform(airquality, new = -Ozone, Temp = (Temp-32)/1.8)

attach(airquality)
# marginally interesting ...
transform(Ozone, logOzone = log(Ozone))
detach(airquality)
```

try — *Try an Expression Allowing Error Recovery*

Description

`try` is a wrapper to run an expression that might fail and allow the user's code to handle error-recovery.

Usage

```
try(expr, silent = FALSE)
```

Arguments

`expr`	an R expression to try.
`silent`	logical: should the report of error messages be suppressed?

Details

`try` evaluates an expression and traps any errors that occur during the evaluation. `try` establishes a handler for errors that uses the default error handling protocol. It also establishes a `tryRestart` restart that can be used by `invokeRestart`.

Value

The value of the expression if `expr` is evaluated without error, but an invisible object of class `"try-error"` containing the error message if it fails. The normal error handling will print the same message unless `options("show.error.messages")` is false or the call includes `silent = TRUE`.

See Also

`options` for setting error handlers and suppressing the printing of error messages; `geterrmessage` for retrieving the last error message. `tryCatch` provides another means of catching and handling errors.

Examples

```
## this example will not work correctly in example(try),
## but it does work correctly if pasted in
options(show.error.messages = FALSE)
try(log("a"))
print(.Last.value)
options(show.error.messages = TRUE)

## alternatively,
print(try(log("a"), TRUE))

## run a simulation, keep only the results that worked.
set.seed(123)
x <- rnorm(50)
doit <- function(x)
{
    x <- sample(x, replace=TRUE)
    if(length(unique(x)) > 30) mean(x)
    else stop("too few unique points")
}
## alternative 1
res <- lapply(1:100, function(i) try(doit(x), TRUE))
## alternative 2
res <- vector("list", 100)
for(i in 1:100) res[[i]] <- try(doit(x), TRUE)
unlist(res[sapply(res,
  function(x) !inherits(x, "try-error"))])
```

`type.convert` *Type Conversion on Character Variables*

Description

Convert a character vector to logical, integer, numeric, complex or factor as appropriate.

Usage

```
type.convert(x, na.strings = "NA", as.is = FALSE, dec = ".")
```

Arguments

`x`	a character vector.
`na.strings`	a vector of strings which are to be interpreted as `NA` values. Blank fields are also considered to be missing values.
`as.is`	logical. See Details.
`dec`	the character to be assumed for decimal points.

Details

This is principally a helper function for `read.table`. Given a character vector, it attempts to convert it to logical, integer, numeric or complex, and failing that converts it to factor unless `as.is = TRUE`. The first type that can accept all the non-missing values is chosen.

Vectors which are entirely missing values are converted to logical, since `NA` is primarily logical.

Value

A vector of the selected class, or a factor.

See Also

`read.table`

`typeof` *The Type of an Object*

Description

`typeof` determines the (R internal) type or storage mode of any object

Usage

```
typeof(x)
```

Arguments

`x`	any R object.

Value

A character string.

See Also

`mode`, `storage.mode`.

Examples

```
typeof(2)
mode(2)
```

unique *Extract Unique Elements*

Description

`unique` returns a vector, data frame or array like `x` but with duplicate elements removed.

Usage

```
unique(x, incomparables = FALSE, ...)

## S3 method for class 'array':
unique(x, incomparables = FALSE, MARGIN = 1, ...)
```

Arguments

`x`	an atomic vector or a data frame or an array.
`incomparables`	a vector of values that cannot be compared. Currently, `FALSE` is the only possible value, meaning that all values can be compared.
`...`	arguments for particular methods.
`MARGIN`	the array margin to be held fixed: a single integer.

Details

This is a generic function with methods for vectors, data frames and arrays (including matrices).

The array method calculates for each element of the dimension specified by `MARGIN` if the remaining dimensions are identical to those for an earlier element (in row-major order). This would most commonly be used to find unique rows (the default) or columns (with `MARGIN = 2`).

Value

An object of the same type of `x`. but if an element is equal to one with a smaller index, it is removed. Dimensions of arrays are not dropped.

References

Becker, R. A., Chambers, J. M. and Wilks, A. R. (1988) *The New S Language*. Wadsworth & Brooks/Cole.

See Also

`duplicated` which gives the indices of duplicated elements.

Examples

```
unique(c(3:5, 11:8, 8 + 0:5))
length(unique(sample(100, 100, replace=TRUE)))
## approximately 100(1 - 1/e) = 63.21

data(iris)
unique(iris)
```

unlink *Delete Files and Directories*

Description

`unlink` deletes the file(s) or directories specified by `x`.

Usage

```
unlink(x, recursive = FALSE)
```

Arguments

`x`	a character vector with the names of the file(s) or directories to be deleted. Wildcards (normally '`*`' and '`?`') are allowed.
`recursive`	logical. Should directories be deleted recursively?

Details

If `recursive = FALSE` directories are not deleted, not even empty ones.

`file.remove` can only remove files, but gives more detailed error information.

Value

The return value of the corresponding system command, `rm -f`, normally `0` for success, `1` for failure. Not deleting a non-existent file is not a failure.

Note

Prior to R version 1.2.0 the default on Unix was `recursive = TRUE`, and on Windows empty directories could be deleted.

References

Becker, R. A., Chambers, J. M. and Wilks, A. R. (1988) *The New S Language*. Wadsworth & Brooks/Cole.

See Also

`file.remove`.

`unlist` *Flatten Lists*

Description

Given a list structure `x`, `unlist` simplifies it to produce a vector which contains all the atomic components which occur in `x`.

Usage

```
unlist(x, recursive = TRUE, use.names = TRUE)
```

Arguments

`x`	A list or vector.
`recursive`	logical. Should unlisting be applied to list components of `x`?
`use.names`	logical. Should names be preserved?

Details

`unlist` is generic: you can write methods to handle specific classes of objects, see InternalMethods.

If `recursive = FALSE`, the function will not recurse beyond the first level items in `x`.

`x` can be a vector, but then `unlist` does nothing useful, not even drop names.

By default, `unlist` tries to retain the naming information present in `x`. If `use.names = FALSE` all naming information is dropped.

Where possible the list elements are coerced to a common mode during the unlisting, and so the result often ends up as a character vector.

A list is a (generic) vector, and the simplified vector might still be a list (and might be unchanged). Non-vector elements of the list (for example language elements such as names, formulas and calls) are not coerced, and so a list containing one or more of these remains a list. (The effect of unlisting an `lm` fit is a list which has individual residuals as components,)

Value

A vector of an appropriate mode to hold the list components.

References

Becker, R. A., Chambers, J. M. and Wilks, A. R. (1988) *The New S Language*. Wadsworth & Brooks/Cole.

See Also

c, as.list.

Examples

```
unlist(options())
unlist(options(), use.names=FALSE)

l.ex <- list(a = list(1:5, LETTERS[1:5]), b = "Z", c = NA)
unlist(l.ex, recursive = FALSE)
unlist(l.ex, recursive = TRUE)

l1 <- list(a="a", b=2, c=pi+2i)
unlist(l1) # a character vector
l2 <- list(a="a", b=as.name("b"), c=pi+2i)
unlist(l2) # remains a list
```

`unname` *Remove 'names' or 'dimnames'*

Description

Remove the `names` or `dimnames` attribute of an R object.

Usage

```
unname(obj, force = FALSE)
```

Arguments

`obj`	the R object which is wanted "nameless".
`force`	logical; if true, the `dimnames` are even removed from `data.frames`. *This argument is currently **experimental** and hence might change!*

Value

Object as `obj` but without `names` or `dimnames`.

Examples

```
## Answering a question on R-help (14 Oct 1999):
col3 <- 750+ 100* rt(1500, df = 3)
breaks <- factor(cut(col3,breaks=360+5*(0:155)))
# The names are larger than the data ...
str(table(breaks))
barplot(unname(table(breaks)), axes= FALSE)
```

`update.packages` *Download Packages from CRAN*

Description

These functions can be used to automatically compare the version numbers of installed packages with the newest available version on CRAN and update outdated packages on the fly.

Usage

```
update.packages(lib.loc = NULL, CRAN = getOption("CRAN"),
                contriburl = contrib.url(CRAN),
                method, instlib = NULL,
                ask=TRUE, available=NULL, destdir=NULL,
                installWithVers=FALSE)

installed.packages(lib.loc = NULL, priority = NULL)
CRAN.packages(CRAN = getOption("CRAN"), method,
              contriburl = contrib.url(CRAN))
old.packages(lib.loc = NULL, CRAN = getOption("CRAN"),
             contriburl = contrib.url(CRAN),
             method, available = NULL)

download.packages(pkgs, destdir, available = NULL,
                  CRAN = getOption("CRAN"),
                  contriburl = contrib.url(CRAN), method)
install.packages(pkgs, lib, CRAN = getOption("CRAN"),
                 contriburl = contrib.url(CRAN),
                 method, available = NULL, destdir = NULL,
                 installWithVers = FALSE)
```

Arguments

`lib.loc` character vector describing the location of R library trees to search through (and update packages therein).

`CRAN` character, the base URL of the CRAN mirror to use, i.e., the URL of a CRAN root such as `"http://cran.r-project.org"` (the default) or its Statlib mirror, `"http://lib.stat.cmu.edu/R/CRAN"`.

`contriburl` URL of the contrib section of CRAN. Use this argument only if your CRAN mirror is incomplete, e.g.,

	because you burned only the contrib section on a CD. Overrides argument `CRAN`.
`method`	Download method, see `download.file`.
`pkgs`	character vector of the short names of packages whose current versions should be downloaded from `CRAN`.
`destdir`	directory where downloaded packages are stored.
`priority`	character vector or `NULL` (default). If non-null, used to select packages; `"high"` is equivalent to `c("base","recommended")`.
`available`	list of packages available at CRAN as returned by `CRAN.packages`.
`lib,instlib`	character string giving the library directory where to install the packages.
`ask`	logical indicating to ask before packages are actually downloaded and installed.
`installWithVers`	If `TRUE`, will install the package such that it can be referenced by package version

Details

`installed.packages` scans the '`DESCRIPTION`' files of each package found along `lib.loc` and returns a list of package names, library paths and version numbers. `CRAN.packages` returns a similar list, but corresponding to packages currently available in the contrib section of CRAN, the comprehensive R archive network. The current list of packages is downloaded over the internet (or copied from a local CRAN mirror). Both functions use `read.dcf` for parsing the description files. `old.packages` compares the two lists and reports installed packages that have newer versions on CRAN.

`download.packages` takes a list of package names and a destination directory, downloads the newest versions of the package sources and saves them in `destdir`. If the list of available packages is not given as argument, it is also directly obtained from CRAN. If CRAN is local, i.e., the URL starts with `"file:"`, then the packages are not downloaded but used directly.

The main function of the bundle is `update.packages`. First a list of all packages found in `lib.loc` is created and compared with the packages available on CRAN. Outdated packages are reported and for each outdated package the user can specify if it should be automatically updated.

If so, the package sources are downloaded from CRAN and installed in the respective library path (or `instlib` if specified) using the R `INSTALL` mechanism.

`install.packages` can be used to install new packages, it takes a vector of package names and a destination library, downloads the packages from CRAN and installs them. If the library is omitted it defaults to the first directory in `.libPaths()`, with a warning if there is more than one.

For `install.packages` and `update.packages`, `destdir` is the directory to which packages will be downloaded. If it is `NULL` (the default) a temporary directory is used, and the user will be given the option of deleting the temporary files once the packages are installed. (They will always be deleted at the end of the R session.)

See Also

See `download.file` for how to handle proxies and other options to monitor file transfers.

`INSTALL`, `REMOVE`, `library`, `.packages`, `read.dcf`

Examples

```
str(ip <- installed.packages(priority = "high"))
ip[, c(1,3:5)]
```

url.show *Display a text URL*

Description

Extension of `file.show` to display text files on a remote server.

Usage

```
url.show(url, title = url, file = tempfile(),
         delete.file = TRUE, method, ...)
```

Arguments

`url`	The URL to read from.
`title`	Title for the browser.
`file`	File to copy to.
`delete.file`	Delete the file afterwards?
`method`	File transfer method: see `download.file`
`...`	Arguments to pass to `file.show`.

See Also

`url`, `file.show`,`download.file`

Examples

```
url.show("http://lib.stat.cmu.edu/datasets/csb/ch3a.txt")
```

UseMethod *Class Methods*

Description

R possesses a simple generic function mechanism which can be used for an object-oriented style of programming. Method despatch takes place based on the class of the first argument to the generic function or on the object supplied as an argument to `UseMethod` or `NextMethod`.

Usage

```
UseMethod(generic, object)
NextMethod(generic = NULL, object = NULL, ...)
```

Arguments

`generic`	a character string naming a function.
`object`	an object whose class will determine the method to be dispatched. Defaults to the first argument of the enclosing function.
`...`	further arguments to be passed to the method.

Details

An R "object" is a data object which has a `class` attribute. A class attribute is a character vector giving the names of the classes which the object "inherits" from. If the object does not have a class attribute, it has an implicit class, `"matrix"`, `"array"` or the result of `mode(x)`.

When a generic function `fun` is applied to an object with class attribute `c("first", "second")`, the system searches for a function called `fun.first` and, if it finds it, applied it to the object. If no such function is found a function called `fun.second` is tried. If no class name produces a suitable function, the function `fun.default` is used.

Function `methods` can be used to find out about the methods for a particular generic function or class.

Now for some obscure details that need to appear somewhere. These comments will be slightly different than those in Appendix A of the White S Book. `UseMethod` creates a "new" function call with arguments matched as they came in to the generic. Any local variables defined before the call to `UseMethod` are retained (unlike S). Any statements after the call to `UseMethod` will not be evaluated as `UseMethod` does not return. `UseMethod` can be called with more than two arguments: a warning will be given and additional arguments ignored. (They are not completely ignored in S.) If it is called with just one argument, the class of the first argument of the enclosing function is used as `object`: unlike S this is the actual argument passed and not the current value of the object of that name.

`NextMethod` invokes the next method (determined by the class). It does this by creating a special call frame for that method. The arguments will be the same in number, order, and name as those to the current method but their values will be promises to evaluate their name in the current method and environment. Any arguments matched to `...` are handled specially. They are passed on as the promise that was supplied as an argument to the current environment. (S does this differently!) If they have been evaluated in the current (or a previous environment) they remain evaluated.

`NextMethod` should not be called except in methods called by `UseMethod`. In particular it will not work inside anonymous calling functions (eg `get("print.ts")(AirPassengers)`).

Name spaces can register methods for generic functions. To support this, `UseMethod` and `NextMethod` search for methods in two places: first in the environment in which the generic function is called, and then in the registration database for the environment in which the generic is defined (typically a name space). So methods for a generic function need to either be available in the environment of the call to the generic, or they must be registered. It does not matter whether they are visible in the environment in which the generic is defined.

Note

This scheme is called *S3* (S version 3). For new projects, it is recommended to use the more flexible and robust *S4* scheme provided in the **methods** package.

The function `.isMethodsDispatchOn()` returns `TRUE` if the S4 method dispatch has been turned on in the evaluator. It is meant for R internal use only.

References

Chambers, J. M. (1992) *Classes and methods: object-oriented programming in S.* Appendix A of *Statistical Models in S* eds J. M. Chambers and T. J. Hastie, Wadsworth & Brooks/Cole.

See Also

`methods`, `class`, `getS3method`

`vector` *Vectors*

Description

`vector` produces a vector of the given length and mode.

`as.vector`, a generic, attempts to coerce its argument into a vector of mode `mode` (the default is to coerce to whichever mode is most convenient). The attributes of `x` are removed.

`is.vector` returns `TRUE` if `x` is a vector (of mode logical, integer, real, complex, character or list if not specified) and `FALSE` otherwise.

Usage

```
vector(mode = "logical", length = 0)
as.vector(x, mode = "any")
is.vector(x, mode = "any")
```

Arguments

`mode`	A character string giving an atomic mode, or `"any"`.
`length`	A non-negative integer specifying the desired length.
`x`	An object.

Details

`is.vector` returns `FALSE` if `x` has any attributes except names. (This is incompatible with S.) On the other hand, `as.vector` removes *all* attributes including names.

Note that factors are *not* vectors; `is.vector` returns `FALSE` and `as.vector` converts to character mode.

Value

For `vector`, a vector of the given length and mode. Logical vector elements are initialized to `FALSE`, numeric vector elements to `0` and character vector elements to `""`.

References

Becker, R. A., Chambers, J. M. and Wilks, A. R. (1988) *The New S Language.* Wadsworth & Brooks/Cole.

See Also

c, is.numeric, is.list, etc.

Examples

```
df <- data.frame(x=1:3, y=5:7)
## Error:
  as.vector(data.frame(x=1:3, y=5:7), mode="numeric")

x <- c(a = 1, b = 2)
is.vector(x)
as.vector(x)
all.equal(x, as.vector(x)) ## FALSE

# All the following are TRUE:
is.list(df)
! is.vector(df)
! is.vector(df, mode="list")

is.vector(list(), mode="list")
is.vector(NULL,   mode="NULL")
```

vignette *View or List Vignettes*

Description

View a specified vignette, or list the available ones.

Usage

```
vignette(topic, package = NULL, lib.loc = NULL)
```

Arguments

`topic`	a character string giving the (base) name of the vignette to view.
`package`	a character vector with the names of packages to search through, or `NULL` in which case *all* available packages in the library trees specified by `lib.loc` are searched.
`lib.loc`	a character vector of directory names of R libraries, or `NULL`. The default value of `NULL` corresponds to all libraries currently known.

Details

Currently, only PDF versions of vignettes can be viewed. The program specified by the `pdfviewer` option is used for this. If several vignettes have PDF versions with base name identical to `topic`, the first one found is used for viewing.

If no topics are given, the available vignettes are listed. The corresponding information is returned in an object of class `"packageIQR"`. The structure of this class is experimental.

Examples

```
## List vignettes in all attached packages
vignette()
## List vignettes in all available packages
vignette(package = .packages(all.available = TRUE))
```

warning *Warning Messages*

Description

Generates a warning message that corresponds to its argument(s) and (optionally) the expression or function from which it was called.

Usage

```
warning(..., call. = TRUE)
suppressWarnings(expr)
```

Arguments

`...`	character vectors (which are pasted together with no separator), a condition object, or `NULL`.
`call.`	logical, indicating if the call should become part of the warning message.
`expr`	expression to evaluate.

Details

The result *depends* on the value of `options("warn")` and on handlers established in the executing code.

`warning` signals a warning condition by (effectively) calling `signalCondition`. If there are no handlers or if all handlers return, then the value of `warn` is used to determine the appropriate action. If `warn` is negative warnings are ignored; if it is zero they are stored and printed after the top–level function has completed; if it is one they are printed as they occur and if it is 2 (or larger) warnings are turned into errors.

If `warn` is zero (the default), a top-level variable `last.warning` is created. It contains the warnings which can be printed via a call to `warnings`.

Warnings will be truncated to `getOption("warning.length")` characters, default 1000.

While the warning is being processed, a `muffleWarning` restart is available. If this restart is invoked with `invokeRestart`, then `warning` returns immediately.

`suppressWarnings` evaluates its expression in a context that ignores all warnings.

References

Becker, R. A., Chambers, J. M. and Wilks, A. R. (1988) *The New S Language.* Wadsworth & Brooks/Cole.

See Also

`stop` for fatal errors, `warnings`, and `options` with argument `warn=`.

Examples

```
testit <- function() warning("testit")
testit() ## shows call
testit <- function()
  warning("problem in testit", call. = FALSE)
testit() ## no call
suppressWarnings(warning("testit"))
```

`warnings` *Print Warning Messages*

Description

`warnings` prints the top-level variable `last.warning` in a pleasing form.

Usage

```
warnings(...)
```

Arguments

`...`	arguments to be passed to `cat`.

References

Becker, R. A., Chambers, J. M. and Wilks, A. R. (1988) *The New S Language.* Wadsworth & Brooks/Cole.

See Also

`warning`.

Examples

```
ow <- options("warn")
for(w in -1:1) {
   options(warn = w); cat("\n warn =",w,"\n")
   for(i in 1:3) { cat(i,"..\n"); m <- matrix(1:7, 3,4) }
}
warnings()
options(ow) # reset
```

`weighted.mean` *Weighted Arithmetic Mean*

Description

Compute a weighted mean of a numeric vector.

Usage

```
weighted.mean(x, w, na.rm=FALSE)
```

Arguments

`x`	a numeric vector containing the values whose mean is to be computed.
`w`	a vector of weights the same length as `x` giving the weights to use for each element of `x`.
`na.rm`	a logical value indicating whether `NA` values in `x` should be stripped before the computation proceeds.

Details

If `w` is missing then all elements of `x` are given the same weight.

Missing values in `w` are not handled.

See Also

`mean`

Examples

```
## GPA from Siegel 1994
wt <- c(5,  5,  4,  1)/15
x <- c(3.7,3.3,3.5,2.8)
xm <- weighted.mean(x,wt)
```

which *Which indices are TRUE?*

Description

Give the TRUE indices of a logical object, allowing for array indices.

Usage

```
which(x, arr.ind = FALSE)
```

Arguments

x	a logical vector or array. NAs are allowed and omitted (treated as if FALSE).
arr.ind	logical; should **array indices** be returned when x is an array?

Value

If arr.ind == FALSE (the default), an integer vector with length equal to sum(x), i.e., to the number of TRUEs in x; Basically, the result is (1:length(x))[x].

If arr.ind == TRUE and x is an array (has a dim attribute), the result is a matrix who's rows each are the indices of one element of x; see Examples below.

Author(s)

Werner Stahel and Peter Holzer, for the array case.

See Also

Logic, which.min for the index of the minimum or maximum, and match for the first index of an element in a vector, i.e., for a scalar a, match(a,x) is equivalent to min(which(x == a)) but much more efficient.

Examples

```
which(LETTERS == "R")
which(ll <- c(TRUE,FALSE,TRUE,NA,FALSE,FALSE,TRUE)) # 1 3 7
names(ll) <- letters[seq(ll)]
```

```
which(ll)
which((1:12)%%2 == 0) # which are even?
str(which(1:10 > 3, arr.ind=TRUE))

( m <- matrix(1:12,3,4) )
which(m %% 3 == 0)
which(m %% 3 == 0, arr.ind=TRUE)
rownames(m) <- paste("Case",1:3, sep="_")
which(m %% 5 == 0, arr.ind=TRUE)

dim(m) <- c(2,2,3); m
which(m %% 3 == 0, arr.ind=FALSE)
which(m %% 3 == 0, arr.ind=TRUE)

vm <- c(m)
# funny thing with length(dim(...)) == 1
dim(vm) <- length(vm)
which(vm %% 3 == 0, arr.ind=TRUE)
```

which.min *Where is the Min() or Max() ?*

Description

Determines the location, i.e., index of the (first) minimum or maximum of a numeric vector.

Usage

```
which.min(x)
which.max(x)
```

Arguments

x	numeric vector, whose `min` or `max` is searched.

Value

an `integer` of length 1 or 0 (if and only if `x` has no non-`NA`s), giving the index of the *first* minimum or maximum respectively of `x`.

If this extremum is unique (or empty), the result is the same (but more efficient) as `which(x == min(x))` or `which(x == max(x))` respectively.

Author(s)

Martin Maechler

See Also

`which`, `max.col`, `max`, etc.

`which.is.max` in package **nnet** differs in breaking ties at random (and having a "fuzz" in the definition of ties).

Examples

```
x <- c(1:4,0:5,11)
which.min(x)
which.max(x)

data(presidents)
presidents[1:30]
range(presidents, na.rm = TRUE)
```

```
which.min(presidents) # 28
which.max(presidents) # 2
```

with *Evaluate an Expression in a Data Environment*

Description

Evaluate an R expression in an environment constructed from data.

Usage

```
with(data, expr, ...)
```

Arguments

`data`	data to use for constructing an environment. For the default method this may be an environment, a list, a data frame, or an integer as in `sys.call`.
`expr`	expression to evaluate.
`...`	arguments to be passed to future methods.

Details

`with` is a generic function that evaluates `expr` in a local environment constructed from `data`. The environment has the caller's environment as its parent. This is useful for simplifying calls to modeling functions.

Note that assignments within `expr` take place in the constructed environment and not in the user's workspace.

See Also

`evalq`, `attach`.

Examples

```
# examples from glm:
library(MASS)
data(anorexia)
with(anorexia, {
    anorex.1 <- glm(Postwt ~ Prewt + Treat + offset(Prewt),
                    family = gaussian)
    summary(anorex.1)
})

with(data.frame(u = c(5,10,15,20,30,40,60,80,100),
```

```
              lot1 = c(118,58,42,35,27,25,21,19,18),
              lot2 = c(69,35,26,21,18,16,13,12,12)),
     list(summary(glm(lot1 ~ log(u), family=Gamma)),
          summary(glm(lot2 ~ log(u), family=Gamma))))

# example from boxplot:
data(ToothGrowth)
with(ToothGrowth, {
  boxplot(len ~ dose, boxwex = 0.25, at = 1:3 - 0.2,
          subset= supp == "VC", col="yellow",
          main="Guinea Pigs' Tooth Growth",
          xlab="Vitamin C dose mg",
          ylab="tooth length", ylim=c(0,35))
  boxplot(len ~ dose, add = TRUE, boxwex = 0.25,
          at = 1:3 + 0.2, subset = supp == "OJ",
          col="orange")
  legend(2, 9, c("Ascorbic acid", "Orange juice"),
           fill = c("yellow", "orange"))
})

# alternate form that avoids subset argument:
with(subset(ToothGrowth, supp == "VC"),
   boxplot(len ~ dose, boxwex = 0.25, at = 1:3 - 0.2,
           col="yellow", main="Guinea Pigs' Tooth Growth",
           xlab="Vitamin C dose mg",
           ylab="tooth length", ylim=c(0,35)))
with(subset(ToothGrowth,  supp == "OJ"),
  boxplot(len ~ dose, add = TRUE, boxwex = 0.25,
          at = 1:3 + 0.2, col="orange"))
legend(2, 9, c("Ascorbic acid", "Orange juice"),
       fill = c("yellow", "orange"))
```

`write` *Write Data to a File*

Description

The data (usually a matrix) `x` is written to file `file`. If `x` is a two-dimensional matrix you need to transpose it to get the columns in `file` the same as those in the internal representation.

Usage

```
write(x, file = "data",
      ncolumns = if(is.character(x)) 1 else 5,
      append = FALSE)
```

Arguments

`x`	the data to be written out.
`file`	A connection, or a character string naming the file to write to. If `""`, print to the standard output connection. If it is `"|cmd"`, the output is piped to the command given by '`cmd`'.
`ncolumns`	the number of columns to write the data in.
`append`	if `TRUE` the data `x` is appended to file `file`.

References

Becker, R. A., Chambers, J. M. and Wilks, A. R. (1988) *The New S Language.* Wadsworth & Brooks/Cole.

See Also

`save` for writing any R objects, `write.table` for data frames, and `scan` for reading data.

Examples

```
# create a 2 by 5 matrix
x <- matrix(1:10,ncol=5)

# the file data contains x, two rows, five cols
# 1 3 5 6 9 will form the first row
write(t(x))
```

```
# the file data now contains the data in x,
# two rows, five cols but the first row is 1 2 3 4 5
write(x)
unlink("data") # tidy up
```

write.table *Data Output*

Description

`write.table` prints its required argument `x` (after converting it to a data frame if it is not one already) to `file`. The entries in each line (row) are separated by the value of `sep`.

Usage

```
write.table(x, file = "", append = FALSE, quote = TRUE,
   sep = " ", eol = "\n", na = "NA", dec = ".",
   row.names = TRUE, col.names = TRUE,
   qmethod = c("escape", "double"))
```

Arguments

`x`	the object to be written, typically a data frame. If not, it is attempted to coerce `x` to a data frame.
`file`	either a character string naming a file or a connection. `""` indicates output to the console.
`append`	logical. If `TRUE`, the output is appended to the file. If `FALSE`, any existing file of the name is destroyed.
`quote`	a logical or a numeric vector. If `TRUE`, any character or factor columns will be surrounded by double quotes. If a numeric vector, its elements are taken as the indices of the variable (columns) to quote. In both cases, row and columns names are quoted if they are written, but not if `quote is FALSE`.
`sep`	the field separator string. Values within each row of `x` are separated by this string.
`eol`	the character(s) to print at the end of each line (row).
`na`	the string to use for missing values in the data.
`dec`	the string to use for decimal points.
`row.names`	either a logical value indicating whether the row names of `x` are to be written along with `x`, or a character vector of row names to be written.
`col.names`	either a logical value indicating whether the column names of `x` are to be written along with `x`, or a character vector of column names to be written.

`qmethod` a character string specifying how to deal with embedded double quote characters when quoting strings. Must be one of `"escape"` (default), in which case the quote character is escaped in C style by a backslash, or `"double"`, in which case it is doubled. You can specify just the initial letter.

Details

Normally there is no column name for a column of row names. If `col.names=NA` a blank column name is added. This can be used to write CSV files for input to spreadsheets.

`write.table` can be slow for data frames with large numbers (hundreds or more) of columns: this is inevitable as each column could be of a different class and so must be handled separately. Function `write.matrix` in package **MASS** may be much more efficient if `x` is a matrix or can be represented in a numeric matrix.

See Also

The "R Data Import/Export" manual.

`read.table`, `write`.

`write.matrix`.

Examples

```
## To write a CSV file for input to Excel one might use
write.table(x, file = "foo.csv", sep = ",", col.names = NA)
## and to read this file back into R one needs
read.table("file.csv", header = TRUE, sep = ",",
           row.names=1)
```

writeLines *Write Lines to a Connection*

Description

Write text lines to a connection.

Usage

```
writeLines(text, con = stdout(), sep = "\n")
```

Arguments

`text`	A character vector
`con`	A connection object or a character string.
`sep`	character. A string to be written to the connection after each line of text.

Details

If the `con` is a character string, the functions call `file` to obtain a file connection which is opened for the duration of the function call.

If the connection is open it is written from its current position. If it is not open, it is opened for the duration of the call and then closed again.

Normally `writeLines` is used with a text connection, and the default separator is converted to the normal separator for that platform (LF on Unix, CRLF on Windows, CR on Classic MacOS). For more control, open a binary connection and specify the precise value you want written to the file in `sep`. For even more control, use `writeChar` on a binary connection.

See Also

`connections`, `writeChar`, `writeBin`, `readLines`, `cat`

xtabs *Cross Tabulation*

Description

Create a contingency table from cross-classifying factors, usually contained in a data frame, using a formula interface.

Usage

```
xtabs(formula = ~., data = parent.frame(), subset,
      na.action, exclude = c(NA, NaN),
      drop.unused.levels = FALSE)
```

Arguments

`formula` a formula object with the cross-classifying variables, separated by +, on the right hand side. Interactions are not allowed. On the left hand side, one may optionally give a vector or a matrix of counts; in the latter case, the columns are interpreted as corresponding to the levels of a variable. This is useful if the data has already been tabulated, see the examples below.

`data` a data frame, list or environment containing the variables to be cross-tabulated.

`subset` an optional vector specifying a subset of observations to be used.

`na.action` a function which indicates what should happen when the data contain `NA`s.

`exclude` a vector of values to be excluded when forming the set of levels of the classifying factors.

`drop.unused.levels`
a logical indicating whether to drop unused levels in the classifying factors. If this is `FALSE` and there are unused levels, the table will contain zero marginals, and a subsequent chi-squared test for independence of the factors will not work.

Details

There is a `summary` method for contingency table objects created by `table` or `xtabs`, which gives basic information and performs a chi-squared test for independence of factors (note that the function `chisq.test` in package **ctest** currently only handles 2-d tables).

If a left hand side is given in `formula`, its entries are simply summed over the cells corresponding to the right hand side; this also works if the lhs does not give counts.

Value

A contingency table in array representation of class `c("xtabs", "table")`, with a `"call"` attribute storing the matched call.

See Also

`table` for "traditional" cross-tabulation, and `as.data.frame.table` which is the inverse operation of `xtabs` (see the `DF` example below).

Examples

```
data(esoph)
## 'esoph' has the frequencies of cases and controls for
## all levels of the variables 'agegp', 'alcgp', and
## 'tobgp'.
xtabs(cbind(ncases, ncontrols) ~ ., data = esoph)
## Output is not really helpful ... flat tables are better:
ftable(xtabs(cbind(ncases, ncontrols) ~ ., data = esoph))
## In particular if we have fewer factors ...
ftable(xtabs(cbind(ncases, ncontrols) ~ agegp, data=esoph))

data(UCBAdmissions)
## This is already a contingency table in array form.
DF <- as.data.frame(UCBAdmissions)
## Now 'DF' is a data frame with a grid of the factors and
## the counts in variable 'Freq'.
DF
## Nice for taking margins ...
xtabs(Freq ~ Gender + Admit, DF)
## And for testing independence ...
summary(xtabs(Freq ~ ., DF))

data(warpbreaks)
```

```
## Create a nice display for the warp break data.
warpbreaks$replicate <- rep(1:9, len = 54)
ftable(xtabs(breaks ~ wool + tension + replicate,
             data = warpbreaks))
```

`zcbind` *Bind Two or More Time Series*

Description

Bind Two or More Time Series which have common frequency.

Usage

```
.cbind.ts(sers, nmsers, dframe = FALSE, union = TRUE)
```

Arguments

`sers`	a list of two or more univariate or multivariate time series, or objects which can coerced to time series.
`nmsers`	a character vector of the same length as `sers` with the names for the time series.
`dframe`	logical; if `TRUE` return the result as a data frame.
`union`	logical; if `TRUE`, act as `ts.union` or `ts.intersect`.

Details

This is an internal function which is not to be called by the user.

`zip.file.extract` *Extract File from a Zip Archive*

Description

This will extract the file named `file` from the zip archive, if possible, and write it in a temporary location.

Usage

```
zip.file.extract(file, zipname = "R.zip")
```

Arguments

`file`	A file name.
`zipname`	The file name of a `zip` archive, including the `".zip"` extension if required.

Details

The method used is selected by `options(unzip=)`. All platforms support an `"internal"` unzip: this is the default under Windows and the fall-back under Unix if no `unzip` program was found during configuration and `R_UNZIPCMD` is not set.

The file will be extracted if it is in the archive and any required `unzip` utility is available. It will probably be extracted to the directory given by `tempdir`, overwriting an existing file of that name.

Value

The name of the original or extracted file. Success is indicated by returning a different name.

Note

The `"internal"` method is very simple, and will not set file dates.

Other Books From The Publisher

Network Theory publishes books about free software under free documentation licenses. Our current catalogue includes the following titles:

- **Comparing and Merging Files with GNU diff and patch** by David MacKenzie, Paul Eggert, and Richard Stallman (ISBN 0-9541617-5-0) $19.95 (£12.95)
- **Version Management with CVS** by Per Cederqvist et al. (ISBN 0-9541617-1-8) $29.95 (£19.95)
- **GNU Bash Reference Manual** by Chet Ramey and Brian Fox (ISBN 0-9541617-7-7) $29.95 (£19.95)
- **An Introduction to R** by W.N. Venables, D.M. Smith and the R Development Core Team (ISBN 0-9541617-4-2) $19.95 (£12.95)
- **GNU Octave Manual** by John W. Eaton (ISBN 0-9541617-2-6) $29.99 (£19.99)
- **GNU Scientific Library Reference Manual - Second Edition** by M. Galassi, J. Davies, J. Theiler, B. Gough, G. Jungman, M. Booth, F. Rossi (ISBN 0-9541617-3-4) $39.99 (£24.99)
- **An Introduction to Python** by Guido van Rossum and Fred L. Drake, Jr. (ISBN 0-9541617-6-9) $19.95 (£12.95)
- **Python Language Reference Manual** by Guido van Rossum and Fred L. Drake, Jr. (ISBN 0-9541617-8-5) $19.95 (£12.95)

All titles are available for order from bookstores worldwide. Sales of the manuals fund the development of more free software and documentation. For details visit the website `http://www.network-theory.co.uk/`

Index

bold numbers denote primary references to sections.
roman numbers denote references to entries within a section.
italic numbers denote references to an entry in the text of another section.

Printed in the United States
33417LVS00004BA/13-15

9 780954 612009